TABLE OF CONTENTS

Acknowledgements

Many people helped me out with this book. Out of the many medical doctors in my life during the time that I was injured, Frank J. Metzger, D. O., was one of the many people that believed in me and provided an endorsement, giving me the credibility to speak about this relationship between epilepsy and osteoporosis. His compassion as a doctor greatly surpasses everybody else who was responsible for my medical recovery.

Lynn Wiese Sneyd, my former neighbor, close friend, and personal foundation throughout this entire process, was there for me providing moral support, in spirit as well as in person, and provided the means with which I eventually learned about how vaccination and disease are connected, giving me the answer to a question that I had for more than twenty years about why I had been diagnosed with epilepsy. As the author of *Holistic Parenting,* her research provided me with the information that helped me understand more about vaccinations and how they are related to a lot of the problems associated with neurological disorders.

I also want to thank Christopher Scanlon, one of my classmates in my high school graduating class who listened to me when nobody else would take my project seriously and for providing me with an opportunity to get my work noticed by an editor at a publishing house.

Carol Postulka, for her support of my projects and for her warmth and compassion when I didn't think that I could make it. Her understanding in my own time of need showed me that there are people out there that understand me to some extent.

Other people that were of indirect help include Charles J. Trimberger, MSW; Marcy Trimberger; Arthur J. Turner, M.D.; Joey Arnel Sayson; David Russell; Kenneth Root, M.D.

Introduction

On a humid summer day in June 2001, I was packing the Ryder moving van and moving 1,850 miles from Wisconsin into the southwest of Arizona. At this time, I had my associate degree in accounting and had no idea how far this was going to take me from the world of education. As I had visited Arizona only two months earlier, I noticed something different within me, something similar to a calming that I had not ever experienced in my life. When I got back home from my vacation, I hit mass depression. I knew that this was something that I wanted to explore. Why was it that I felt better anytime that I visited Arizona? I had no idea how far this would take me from conventional medicine and make me challenge the role that education plays in this country. I had no clue that I was going to be working in something completely different than what I had my degree in.

The Epilepsy and Osteoporosis Link is more than the link that exists between epilepsy and osteoporosis from a medical perspective. This book explores how the process of change can result in euphoria while at the same time tear apart a family. Finding that the changes that I was undergoing intellectually were not shared directly by my family, it became apparent what I needed to do in order to protect my vision. I had to cut off all communication with my family and focus on the task at hand. How could I even relate to my family what changes I had gone through? Whenever I am around at Thanksgiving and Christmas, my parents don't ask me at all about my projects. I am always someone who tries to learn new things – even if I make mistakes when I try things that have never been tried before. After all, if I don't make mistakes, then that really means that I am not learning. I only make a mistake once, however.

After I was diagnosed with secondary osteoporosis at twenty-seven years of age, in April 2002, due to my bone density being fifty percent below normal, I had to give up my inline skating. When my rollerblades started to gather dust as they sat on the shelf in my closet, all I knew was that I wanted to get better. As athletic as I was, I did not want to risk any fracture occurring. After I checked myself in at the hospital, I got referred to an osteopath, an internist, an endocrinologist, a neurologist, and a urologist. This life altering experience would be what would cause me to challenge the fabric of the American Medical Association.

At the time, doctors were slowly finding out about the link between epilepsy and osteoporosis. For a long time, osteoporosis was something that was found to be a disease that affected women who were past the age of menopause. The earliest studies that had been done on patients with epilepsy were those who were institutionalized and on the early forms of medication. As the medication got better and allowed people to live a better quality of life, there was an odd factor showing up. Children who were taking these anticonvulsants did not have any effect on their bone density. The drug companies knew that this was a side effect and thought that it was not a serious issue. After all, doesn't everything in this world have side effects?

When I found out that Arizona did not have a high risk program for people who could not get insurance through normal means, I thought that I would be working in an accounting job, which would provide me benefits. When that ended up not working out as planned, I did everything that I could to help out with financial expenses. The one job that I did take would end up saving millions of lives. After my insurance got cancelled because I was no longer a resident of Wisconsin, I started to look at holistic medicine with the idea that this is the reason that I have epilepsy. I had to start making these assumptions. After all, I did not want to be on medications that were causing more and more problems. What would happen after this, would be the one thing that would spell job security. In 2003, I never expected that this would lead me to take a trip to Los Angeles to meet with a producer to do a script consultation and meet with an entertainment attorney to learn more about Hollywood in general. I had a movie script pretty much completed. As a first time writer, it was difficult to get people to listen to my message.

My ambition would end up almost costing me my life and freedom on many occasions. Ever since I started down this rocky path, I had people who were giving me the one-fingered salute out their car window. I had succeeded at doing one thing so far, which was isolating myself – which was for my personal security – and making those around me very angry because I was using my difference to educate people on something that was, at the time of my surgery, just coming down the newswire. My father absolutely disagreed that vaccinations were the reason that I had epilepsy. There were people out there that wanted to murder me. When my bodyguard had to shove me out of the way of a car that had the intention of running me over, I realized I was in serious peril. As I brought this situation up to my parents, they were insensitive, saying that I had changed. I toyed with the idea of whether or not it was going to eventually be necessary to carry a weapon. I was not someone who was inclined to think in this manner, but it was as though the world did not want to know of my presence.

When I had a number of incidents come up during the time I was taking classes to continue my four year degree, I dropped out of school after I was hated on sight by a couple of teachers – and especially by those who found out about the fact that I had received a large insurance settlement for my injuries. Especially when I had a number of close calls with people jumping up and down in front of my car while I was at a stop light, I realized that if I wanted to have anything resembling a normal life, I was going to have to hire security – and drop out of school quickly before I lost my mind. It was around this time that I realized that all education does is dumb people down and does not encourage creativity by any stretch of the imagination. All education does is turn people into robots who are trained to do nothing except follow orders. I decided that this was not the kind of life that I wanted to have for myself any longer and filed for a divorce from the educational system that I was enrolled in.

I would find on many occasions that when I went to interview for jobs, that the word was getting around like wildfire about my story and I couldn't get a job anywhere, because I was "different". Being out of work for over a year while I was healing from my hip injury didn't help either. Whenever employers would bring up the proverbial phrase to "tell me about yourself", I felt that I had no choice but to bring up my health information, which is not usually supposed to be disclosed in the interview process. What I found in many situations was that employers lacked the ability to be compassionate toward my situation. I didn't know how I would be able to get them to understand what it is like to be in the prime of your life and have a hip fracture. It wasn't my fault that I took a job that would teach me something that I otherwise would never have found out about.

Ever since I had been through this life altering event, I never thought that I would be challenging the entire educational system and how they promote conformity with values ingrained in society instead of letting people be themselves. It is this same institution that promotes the standards of the American Medical Association, which says that any doctor that does not practice within the confines of its boundaries are considered to be quacks. Realizing that having a college degree was going to result in nothing but dead end and routine jobs, I decided that I wanted something more than a nine to five job. The problem, was that people have no idea how long it takes to get a movie contract, or for that matter, to sell your screenplay with tens of millions of dollars on the line. Under the assumption that I would not be there for the long-term, I end up not getting called back for a second interview.

As I started to grow frustrated with the entire situation, I felt that I had no choice but to self-publish this book so I could at least bring some money back into my bank account. I realized very quickly that I couldn't wait around for a "real" literary agent to pick up the phone and give me a call. As this book almost never got published due to the work of a scoundrel, that experience opened my eyes to the subtleties of the publishing industry that I was not ever aware of until I started writing my book. As a result, I started writing *Euphoria*, a movie that is based on my experience as a survivor of secondary osteoporosis. If it were not for this agent, my book would probably be like every book that sits in the bookstore. I needed people to ask for my book.

Since a lot of literary agents were passing on my proposal because of the fact that I lacked the medical background to talk about that which doctors clearly had no clue about, that brought up another question. How could doctors be considered an authority when they were just as confused as everyone else? It was like the chicken before the egg. As a result, I felt that I had no choice but to consider becoming an educator with regard to this rare but very serious medical mystery. When Trimberger Enterprises was born, it was just this book. However, I was going to find that the book was going to have to work in conjunction with the movie in order to turn me into that teacher. Unfortunately, no agent was interested in taking me on as a client, since there is always that catch-22 situation with regard to publishing. No publisher will take authors that don't have an agent, and no agents will take you if you don't have a publisher interested in your material. It was at a ten year reunion that I met somebody who may be able to find this book a home within a publishing house and sent him two sample chapters from my book.

In the wake of a studio passing on my script because they said that there was not enough of an adequate market that would make my movie commercial, my jaw dropped. I had the facts about how many people were visiting my site. I even knew where they were located. How could they say this without talking to me? My idea was high concept. It was never tried before. The problem, was that I could not get financing to get the ball rolling on this. There were even times that I was asked by producers if this, indeed, was a true story. When I made it clear that my story was true, it was difficult to get people to listen. At the time, I did not have a Hollywood track record and was starting to go out of my mind trying to get somebody to listen to me.

Turning to talent, I started a letter writing campaign to the agencies of some bankable characters to see if they would get behind the script. Specifically, after seeing Leonardo DiCaprio in the movie *Catch Me If You Can*, an inspirational movie with a surprise ending, I realized what I needed to do. Turning into a movie buff overnight, I started to watch movies with a different stance, now that I had a completed screenplay that I am trying to sell. Every movie that I went to see, I analyzed the movie and tried to figure out where the breaks occurred between the first, second, and third acts so that I would know how to write my screenplay properly.

This book was self-published to serve as the architectural blue print for proving that there is a viable market that is interested in my story. My web site statistics even show that there are huge numbers of people on the east and west coast that are interested in my story.

In chapter five, I talk at length about the process that I had to go about learning about Hollywood, especially since I am the only one in my family who has taught myself about how to write a movie script – all without a formal education. It all started with a book that I bought about how to put a screenplay together. While that sat on my shelf, I never thought that I would be taking that road. As time went on, I purchased more books on entertainment law as it relates to screenwriting. It was around the time that I sat in the hospital awaiting surgery that I knew that my life was going to change for the better. Unfortunately, there were obstacles that I would have to jump over in order to get that to work with any degree of success. It seemed that those in my immediate family did not want to do anything to help me to succeed. They just wanted me to get a job. Jobs are boring, at least for those who have a story to share with the world or for those who are trying to change the world.

It seemed like my parents were pulling out all the stops to get me to listen to them. They were concerned that this was all that I was caring about. I felt a moral obligation to get the message out to the world. Their life did not almost end prematurely as a result of being on a medication that prevented vitamin D from being produced by the body. It was this situation, compounded by the loss of my insurance in 2003, that would take me far outside the realm of what I considered home. My parents were clearly upset with how much the projects that I was working on consumed my life, upset with how driven I was to make something happen. I had been sacrificing sleep to get a lot of this work done. I shared nothing in common with my parents any longer. Anything that came out of my mouth was in regard to my projects.

Sometimes the best teachers are those without any formal training in school. Life experience is what leads the way for me. Through this life experience, I attempt to show you the way as well.

Chapter 1

Living Life in a Blue Dream

It all started with an out of court settlement for $10,000 – and it was in no way going to end there. As a new author, I was told, not only are the advances paid to authors not substantial, but that a lot of times authors have a hard time generating publicity for their projects. Not wanting my book with my life story to go out of print, I do something that makes my problems worse. I draw attention to myself until it gets to the point that my life is no longer normal and I end up crossing a fine line. This all started with getting hooked up with a literary agent that was not being 100 percent truthful to me.

After my literary agent was shut down by San Angelo Police, not only was I without representation – I had a difficult time getting an agent. I was fed nothing but lies by this agent. Then I realized something so horrible that I almost choked. That was also when I realized that no agent in their right mind was going to try and sell the manuscript that had been seized into evidence. I helplessly watched the hits to my website take a huge dive and had to do something immediately to revive my image and save myself. I started over the next few months as the hits to my website were slipping toward ground zero, to write a movie off of my experience. That is when Euphoria was born, a project which would strand me economically. Not only was I unable to gain representation from an agent or producer, but I was also unable to get a job. I am guessing that this has to do with the fact that I have overqualified myself based on my experiences, not to mention that I would not have staying power if Hollywood came along tomorrow and offered me a million dollars for the copyright to my movie script.

At the same time as I realized that my movie was not going to be produced due to no market I realize why this could be the case. My book has not been released. So, I make it my goal to get my book directly into Lenoardo DiCaprio's hands. As I was starting to draft the new and updated version of The Epilepsy and Osteoporosis Link, which is the form that you are reading currently, I realize I have just solved the publicity problem. When I first started writing my book, I was a nobody. Before much longer, my name will be in the trades – and I will probably be booked on The Tonight Show to speak of my experience somewhere along the line, assuming that I have permission by the producer to give interviews. I found that I had just solved one problem and created a complete new problem that I could do nothing about.

With the full knowledge that people might scheme to plot against me somehow, until further notice, I had no choice for my own security to treat everybody as suspect. I didn't like the idea that I would have to do this. On the one hand, I wanted people to know me more, but I wanted to maintain some level of civility.

BROKEN BARRIERS

I felt like I was about to jump right out of my skin. The euphoria building up inside of me was so strong that it was knocking down walls, and not just the walls of personal security. I had a lot of walls stacked against me. There were people who read my initial screenplay, who were finding one thing after another wrong with it. They told me that it would never make it through the studio and into the spec market. I decided to prove them wrong out of anger. After all, coverage is just one person's opinion of the material. There was simply no way that I was going to get a thumbs up from this particular script coverage company, who I will be nice enough not to put in print. I believe that part of it had to do with my story, which was unique in its own sense of the word. Somebody out there does not want me to set a fire, and they are making it damn hard to find a book of matches.

I was only going to start trading one set of worries for another set of worries. I realized that I had a fire started that was roaring out of control. People were reacting to me differently. The friends that I used to have stopped responding to me in the same way. My own family, I felt, had turned against me in a sick, demeaning way. I was literally like a small island in a big ocean. With no emotional connections, nobody to tell me that I was doing things right, it is any wonder that things got this far. On the one hand, my parents wanted me to succeed since they were paying my rent. On the other hand, they did not want me to be so far away from home, and they had no way of understanding why I was isolating myself.

After mentioning to them that there was a possibility that I would be moving to Los Angeles from Phoenix to pursue a career in screenwriting, they want me to stay where I am. Because once I cross that line into the realm of celebrity and fame, there is no way to turn back in my life. I already set that fire a long time ago. It all started during the time that I took a job that I normally would not take.

Summarily told to get a regular job by a lot of people instead of focus so much energy on hoping things work out over the long term, I start to grow angrier with the people saying this. It was as if the people who were saying this did not know any other way to do things. I knew that there was a glimmer of hope. My horoscope predicted it.

After all, I am a Gemini, and every day seemed to predict that something would happen. For me, those days never seemed to come. That is only the beginning of the war.

I used to be getting passing grades, but with the chance that my movie script might sell, I dropped out of school, after finding that my grades start to slip lower and lower. I was not as emotionally invested in my schoolwork as I used to be. In addition, I felt like certain teachers were treating me unfairly.

In addition to that, there was a vibe of anger that I was sensing among the people who I thought I could call my "friends". It all started one day, when I was sitting inside of a restaurant, and someone snuck up behind me to get a closer look at me. I didn't know when this day was going to come when I would cross the line between an ordinary person and that of a famous person or celebrity but one this is certain. I was prepared to get into a fight if necessary for self-defense – and for the record, I *never* start fights. I only finish them.

I didn't see any reason to emotionally invest myself in school after getting my script requested only an hour after blasting it all over Hollywood. However, it would be a long journey and I was started to get impatient. What was never told to me was when there were tens of millions of dollars on the line, that things never move as quick as I wanted them to move, especially when you are an unproduced screenwriter. Nobody comes looking for you – you have to go to them and give them a reason.

Over the past few months, I notice that there have been a lot of weird things happening. I had been living an otherwise ordinary life, but word of my story – and my internet presence – was making its way around school like I had just started a wildfire. This is not the type of fire that firefighters are going to be able to extinguish. Most fires burn out on their own around Arizona, but not this fire. Throwing kerosene on the fire that I started, I ended up making it an out of control wildfire that was about to engulf me in its flames. Unfortunately for me, however, this fire would not be strong enough to burn anything. It was a fire that ultimately died with no way to rekindle it. My plans were to get this book published after I got option money for my movie, which never came along. With my mother losing her job, and with no hope for me in regard to getting employment by a studio or a producer anytime soon, I was left with no options. I had shot myself in the feet and had no way to walk away from this train wreck.

My success would eventually work against me and create a personal security problem that I was not equipped to deal with. Completely attacking the core of who I was as a person, never in a million years, would I ever expect that the possibility existed that I would have to hire security in certain situations if I ever wanted to resume some level of a normal lifestyle. With no way to pay security on a regular basis, I did with what I had. If that doesn't yet tell you about why I was so unique in my own way, you will understand as you read through the first part of this book, particularly Chapter six where I talk about *The Hidden Price of Fame.* It's nothing to laugh at or disregard. Take it very seriously.

My life was starting to change at a pace that was quicker than I was used to dealing with. It was like I was outgrowing my own body as I slowly got to an elevated social and financial status. As I write this, I am still waiting for a miracle to happen somewhere. I sit here, going through who I am going to have to hire to manage my day to day affairs so I can deal with the chaos that is erupting out in the world. I certainly don't want to have a repeat of dealing with a literary agent that was a shark and only after my money and fed me nothing but empty promises about my abilities as an author. My life as a writer would be one that was unlike that of any other writer's experience. I had a story to tell – but was unable to tell it because I did not understand enough about myself – until two years ago.

THE MARKETING PROBLEM

From Hollywood's point of view, their attitude is, "Why should anybody care about my story?" I was put in the position of having to educate people on the link between epilepsy and osteoporosis, something that for the first time in my life, I had no idea how to do. How do you describe what it is like for intense pain to shoot up your leg as a result of fracturing your hip? How do you explain that because the work has never been done in a theater before, that people are going to want to see this movie made? As a result of this, I had no idea how I was going to handle this.

There were so many barriers to entry that it was difficult to get anybody to listen. Literary agents who were legitimate often questioned my authority to be able to write something that was about the link between epilepsy and osteoporosis, saying that I needed to have a medical background in order to do this. Hollywood was saying that my idea was simply not commercial. My parents were threatening me and getting very aggressive telling me to get a job. After putting up a website to market my work that I was doing, it never occurred to me that this was going to come back and smack me in the face. I grew tired of knowing that certain people, for one reason or another, did not want to see me succeed.

Other people around me were not accepting of the fact that I was different The job that I thought I would have that would pay for my medication and would give me stable employment for many years, was ripped right out from under me, before my insurance went into effect after my employer found out that I had epilepsy. Employers did not want to hire me for fear that I was going to bolt within six months or whenever my story sold.

I was almost at my wits end and was thinking about carrying a weapon to protect my own interests. I didn't want to arm myself because I would definitely not be understood as to why I would do something like that. To understand why I say this, you will get a better idea in Chapter six where I talk about the price of fame. I was fearing that I was going to walk outside one day, and find that my car was no longer where I parked it. ***Welcome to my world.***

THE ONE JOB THAT CHANGED EVERYTHING

If it weren't for the one job that I would not normally take, I would be living a normal day to day existence. I was in a hurry to get my story told. If it were not for this, my book would just be like every book in the book store – different. I didn't want something to be different. I wanted something that would start a fire. But getting that fire started, as I would find, would prove to be a very difficult task. First it had to do with production companies saying that there was no market for my screenplay. Was I hearing this correctly? Everybody I talk to had no knowledge about the role that anticonvulsant medications played in connection to low bone density.

After finding out that my insurance had been cancelled after a major operation, a new seed, Trimberger Enterprises, was born to try and fight the evil that lurked in the world. With no way to pay for expensive prescription medication, unless I chose to live the rest of my life below the poverty level with my parents paying my expenses every month, I needed a way to make a very big splash in this world to buy some time, like say, another one to two decades so I could afford to spend the time growing my business instead of slaving away at a job that I really couldn't give a shit about. With my mother paying my rent for the better part of three years while I struggled to get things under wraps, I needed a miracle – badly.

It was a trip out to California to do a script consultation with a working producer in Hollywood that would eventually turn the table for me. But first, I would almost die as I started to run a severely high fever. After taking a job that I normally would never take, I find out the extent to which the prescription medication that I had taken for fifteen years in my childhood was going to make me consider teaching. I never thought about teachers as those who would make a lot of money.

On September 4, 2004, I anxiously started to watch my email with the hope that I would open my email and find some update from the production studio that had possession of my movie script. What they should have been doing is calling me on the phone. Why am I bothering to check my email? For three months now, they have had the script, and so far, they have not said no. They also have not said yes. I was told by my attorney, that if they wanted to option the script, they wouldn't wait around, sitting on their laurels. I would have gotten a phone call from somebody at the production company immediately. I was starting to get a hint that maybe they were going to option my movie for pre-production, but my email has been quiet.

My profile on inktip.com, for the past month and a half, had not had any visitors. Now I am really starting to jump out of my skin. In order to understand what led me into the world of screenwriting from that of being a serious business student at school, you have to start from the beginning of my life story to fully appreciate how I have developed over time. Only then can you appreciate why it is that I decided to drop out of school and become a teacher.

THE ADVENT OF DETERMINATION

When I was born, everything was normal. I was given the usual round of vaccinations that every child in that hospital would get. At the time I was two years old, my parents bring me back to the hospital because I had for some unknown reason, had a febrile convulsion. Very often, during my childhood, I would complain about the fact that

I felt sick. I would find out why I had this convulsion decades later, when I completely threw caution to the winds and took a two thousand mile road trip all the way across the country to visit my former neighbors, after they moved to Tucson, Arizona. They had just been through some difficulties of their own as they found a newer way to do things. How anybody can stand the heat in this state is beyond the scope of the imagination…I think in a two year period of time my skin color went three shades darker than it was originally.

The pediatrician prescribed Phenobarbitol, a medication often prescribed when someone first starts to have seizures, with the hope that I would grow out of this. Fate was not going to be on my side. My father said that people who have febrile convulsions as young children very often get diagnosed with epilepsy, according to what my pediatrician told him. These thunderstorms that rendered me unconscious would be a part of my personality for the next two decades until I found a way to control them. During this time, I was on numerous medications, which affected my appetite to the point where my desire for food was non-existent. I would often retreat to the bedroom after becoming sick to my stomach at dinner.

After doctors had wires hooked up to my head and witnessed a seizure occur, I was diagnosed with temporal lobe epilepsy. Everything between the time I was diagnosed with epilepsy and the day that I had my hip fracture on me in three places is pretty much a blur. Anytime that I would ask anyone who has a family member with epilepsy how it is that they got, the most common answer was that nobody knows why. The doctors gave me an answer, saying that my epilepsy was idiopathic, meaning that there is no known reason for why it happened. As young as I was, I was not dumb enough to believe this. If those around me thought that I would be taken for some sort of idiot, they were to be sadly mistaken. In the process of trying to get to the truth, I would have no choice but to end up making some enemies, which was not my style. Unfortunately, I can't make everybody happy.

The next two decades would be spent with me asking my parents why I got epilepsy. They did not have an answer for me – only that my nervous system was very fine tuned. With no other information to go and compare this to, I had to depend on my parents to make all my medical decisions. At four years of age, after I was off the Phenobarbitol for a period of time, I had another seizure. The pediatrician prescribed Tegretol, a medication that is used for people with generalized seizures. This medication also had to be taken in very high doses in order to be effective in controlling seizures and would be instrumental in telling me that I am different. Not only was this medication not adequate in controlling my seizures, it also was instrumental in changing my life – and millions of people's lives – for the better.

As time went on, I noticed that my left heel that would be painful to walk on every once in a while. At the time when I was taking medication, bone density and bone scans were not part of my medical ritual. Doctors were not even aware of the fact that anticonvulsant medications were directly related to losses in bone density after adolescence. It would take someone with higher education to teach that to them. Somebody with life experience who would end up being a survivor of secondary osteoporosis. That person, would be me.

DEJA VU

There was one time that I was actually awake during the time that I was in a seizure. And it was this time that really scared the life out of me. What I remember, is that my left hand was developing a mind of its own, and it was as though I had no control over its movements. It was as if my body was possessed by the demon – an evil spirit who was able to control every move my body made. Resisting this evil spirit was not an option. The devil child would end up making the wires short circuit on a regular basis. With the wires in the computer short circuiting all the time, I had to find a way to fit in with the rest of the people my age. And this was not going to happen for any one of many different reasons.

A lot of my times during my high school years, my time was spent in the nurse's office for several hours recovering from migraine headaches that happened as a result of the grand mal seizures. It seemed like my life was unnecessarily interrupted for no good reason. Often, my mother would have to leave work early to come and pick me up at school. There were a lot of times that a seizure would interrupt my day when I was at school. One thing about being in the nurse's office that I vividly remember, was when I was being weaned off of one of my medications. I was on so many different medications at different time intervals that I don't know what I was on during what time period. But one thing is for sure. As a result of being weaned off this medication, I was shaking beyond my own control. And there was nothing that I could do to stop the shaking. It was as if my brain had a mind of its own. The shaking was similar to the freezing climate that I was living in, not knowing that my epilepsy was influenced by it for the most part.

SIMILAR DIFFERENCES

We all grow up comparing ourselves to one another. If a person is different, for any reason, that person is singled out or excluded. We are all the same – we are different from one another – although I was going to redefine different when I moved to Arizona. In my case, I had to deal with something that made me different that I could not even describe to people. The one thing that would make me different would puzzle every neurologist in this country. It would take the right circumstances to bring to light the dangers of anticonvulsant medication.

Having temporal lobe epilepsy meant that, during the time that I was unconscious, I would do things when I was unable to describe to those who witnessed the occurrence. I had to rely on the account of people who knew me well enough to tell me what happened. Because I had no way to describe to people what was happening, I intentionally built a wall around myself, making it difficult for those to be able to understand me. Since I had no way of telling people what my seizures were like, I was at the mercy of my mother to provide the vivid details. If people thought it was difficult to understand who I was as a person with grand mal epilepsy, give it twenty years. Then you will be scratching your head a lot more asking yourself who in the hell you are dealing with.

It would be this same difference that would break that wall down and everybody walking the planet would eventually know who I was if they didn't already know. At the same time that I needed friends, I didn't want to have *this* many friends. Valuing my personal privacy, without knowing what the hell I am doing, I ditch my landline phone. Two years later, when I start talking about my story, putting the words to paper, I never stop to think about how much I have changed. Now, with everybody and their dog knowing me, I am glad that the landline phone is gone. At least that buys me a certain level of protection from the crazy madness out in this world that is intent on hurting those of us that try and get change started.

BUZZ

What I never knew, was that the same thing that made me different, would be the one thing that gave me a voice and would generate enough buzz so that everybody knew who I was. At the time, I did not realize that. I only saw the fact that I was unable to get a driver's license until later in life. If I thought my problems were bad when I couldn't drive due to being different, there was nothing I could do about the fact that I was actually losing friends for being different. Everywhere I went, people were talking about me. I had enough determination, after the near failure of my freelance startup, to find another way to get the message out to everyone about what happened to me on that fateful day of April 2, 2002.

The amount of determination that I had scared me. I was told by a company that covered my script, but kept passing on it, that even though my script was full of incident, it lacked the drama that would make a person care about my situation. Were these people insane?

After getting that round of notes, I took my script and blasted it all over Hollywood, as a way to show them that they were full of shit. Every time I would change something, there was something else wrong with the script. I saw this as nothing but a money laundering business. Angry at these notes, I blasted the very same script all over Hollywood – and one hour later, I had a potential buyer. Unfortunately, they would also be the one to tell me that there was no market for my material. Was I hearing things? How could this be? Everybody I talk to has amazement that I suffered through something so horrific. I found out part of the answer to why there was possibly no market for my material. It had to do with the fact that a lot of the research done with regard to the connection between epilepsy and osteoporosis was done outside of the United States, which is why doctors were unable to understand how to properly diagnose my condition. Because nobody was able to understand me completely, I was left with no choice but to find a way to get my message through. This would not prove easy.

I never thought that I would find myself studying the process of putting a screenplay together and committing myself to this project, pretty much ignoring the rest of my life as I attempted to get paid for all this work. A lot of people who work at screenwriting probably toil at it for many years trying to write a screenplay that can be bought. There are classes that teach people about screenwriting. I don't believe in them.

Writing is simply something that I feel cannot be taught. You are either a writer deep down, or you are not. It takes the right set of circumstances to make the words come out and flow onto paper like an uncontrollable river. Then something came out of left field and completely shocked me. I had been living in Arizona for three years and realized that I had changed.

My own family, who I thought would be supportive emotionally, did not seem to understand me anymore. Had I changed that much? I guess I must have changed somewhere – I had been sacrificing my sleep, and not keeping in regular contact any longer, as my goals were not in compatibility with what my parents wanted out of me. I had one goal – and one care – in this world. I planned to sell this screenplay. At the expense of relationships with my family and friends, this was priority. I felt a moral obligation to somehow get the message out. The problem – nobody would allow me to run my mouth and tell them what they were doing wrong.

I was doing something wrong, however. I wasn't getting any calls back requesting the script. I must have written over a hundred letters to agents, only to find that I shouldn't be contacting agents. At this point in time, I didn't need an agent. I needed a producer. Producers like to work with new people, as they are looking for new material to develop on a regular basis. That would land me an agent later on. But, I was in a hurry and was led to believe that agents would take me as a client.

I had to deal with an overprotective mother, who would not let me run my own life, who would go to all lengths to remind me of things that I already knew. Although what I would find as I got to know myself better – and found out that I should be doing something different – that she was the one who was the most understanding of the situation I was in to a small extent.

I couldn't drive at sixteen like everybody else. I picked up a little idea and started to rollerblade down by the lakefront. I was exercising on a regular basis building up strength in my lower body. Plus, the entire time, it was like a rush being the epicenter of destruction around hundreds of people who were sunning themselves on the beach. My primary way of getting around until the day came when I could qualify for a driver's license, I went rollerblading about every other day. One of these days, I will resume that habit. I haven't rollerbladed for over two years for more reasons than one.

I was excited when I was finally able to get my learner's permit around the time that I was starting to go to college. Then I hit the wall – again. The stress from taking a full time load at the same time as figuring out how to make enough money so that I could pay the bills from month to month was starting to haunt me. I had a seizure and had to turn in my permit. This put off any plans that I would have to spread my wings. At this point, I had no idea that I would be taking a two thousand mile road trip to Arizona to teach the world a lesson.

Somehow, all my problems were going to change. Like around the time in 1996 that I got a call from my neighbor, Lynn Wiese, telling me that she was moving to Arizona. I did not know this, but a little voice was telling me that I was going to be following in her footsteps. She never had mentioned anything about Arizona – until the day I was living with my brother. My mom told me that Lynn Wiese, my next door neighbor, had something that she wanted to tell me. For now, however, I put myself in denial and knew that I had things good where I was.

What she had to tell me would end up changing my life for the better. She told me that she was moving to Tucson, Arizona because her husband's health was difficult to deal with in the cold Wisconsin winters.

I chose not to act on the commands from this voice in my head. I was like a person stuck in a rival dictatorship. I was in a mode where I was working and trying to get my AA degree in accounting. I spent all that time going to school to get my associates degree in accounting – only to find that there was another message that had to be told to the world. This would last five years.

Shortly after getting my license, I made a decision that was going to change the world – without me even knowing it. Somehow, I managed to drive two thousand miles down to Tucson, Arizona from Wisconsin. Getting down to Arizona was a rush. It would also redefine me at the core of my personality, putting me at odds with my parents who did not want me to succeed even though they verbally said they were happy for me.

The trip to Arizona redefined my personality, and I sensed the feeling that I was being controlled by a higher power. As my car traveled through Grand Junction, Colorado and across the Rocky Mountains, I saw something that was a motif to my development. A scenic sunset pointed the way directly into paradise for me. My vacation to visit my former neighbors in Tucson would end up turning me into something that I never thought I could do – teach – about the fact that there is a confirmed link between epilepsy and osteoporosis through the long term use of anticonvulsant medication.

It would also be the one thing that would give holistic medicine a newer meaning that it did not have for my life. I was convinced that my medication, combined with the non-turbulent barometric pressure system, would be enough to help prevent my seizures. When I was writing this book originally, I felt that since I did not have a medical background, I could not speak about epilepsy and osteoporosis. However, my circumstances of being a survivor of secondary osteoporosis gives me the license to speak about this disorder and why it happened to me the way it did.

I was desperate to get out of my parents' house. Living with my brother on the east side of Milwaukee, I found that something still was not there for me. Living with my brother got me out of my parents' domain. It was as though I was not meeting anybody socially. Although I enjoyed living with my brother, I wanted to be closer to school because I did not have access to a car at the time. This was the start of my journey to get my wings spread.

SCARED FOR MY LIFE

In the summer of 1998, I moved out of the upper flat that I shared with my brother, intent to start my own life somehow. That is when I ran into the roommate from hell. The only other person that I knew was a roommate who lived with my brother and his friend for a year. The person who seemed like a nice person that I had seen around the campus, was actually nothing but a jerk with an explosive temper. When I visited this apartment, all of the roommates were moving out at around the same time. At this point, I was not suspicious – people move around a lot in their lifetime. I made the decision to move my belongings into this upper flat. Then he dropped the bombshell on me and I found out that he had a live in girlfriend – after I had moved my belongings into the apartment. She had full rein of the place, and felt like she could just say things to me, and act like a loyal follower by standing up for this roommate. There were times when I found things missing – important things – like a textbook for one of my classes, which he gave to me, saying that another roommate had borrowed it. I knew that this was a bunch of bull. My seizure activity was starting to get out of hand as a result of wondering what he had planned for me.

. This not only caused my stress level to increase, but it also caused my seizure activity to skyrocket. This experience was teaching me one thing - that I could not trust people. My goal was to stand my ground with this bully and put him in his place. I knew that if I left, he would win. On the other hand, if I didn't leave, I would be so stressed, that my seizures would not be under control. I wasn't sure if I was going to be moving. But I had to find another roommate to replace the one that moved out the prior month. The person who moved in was much taller than this bully. Hmm. Good plan. I will find a way to get him in here, and that is when I will make my move. I'll just make his life a living hell while I have the chance – and then make a run for it.

When I had to put a lock on my bedroom door, it was not too long until I decided to strategically plan my escape from this hellhole. Since I knew what his patterns were, I took the liberty of waiting for the house to be empty before packing my stuff into a moving truck and moving to my new location – without telling him. I moved out of there without even bothering to leave a forwarding address. It was almost like a race against time.

Then I realized something else. In the midst of doing all my organizing, I had left phone bills out on the table in the dining room and suspected that he had gotten the number off there, which could easily be cross-referenced to find out where I was living. I called up the phone company and requested that my number be changed to an unpublished number, as I feared he would use the lines of communication to terrorize me. This was worse than a stalking situation. Later on, when I had to return to him to give him the key to the apartment, I walk up to the house and notice that the light in my room was on. He was probably surprised that the room would be devoid of any human existence when he got home. However, he had his way of using psychological manipulation with me.

After I moved into my new apartment, there was a knock on my door from the landlord who gave me a note. It was from my roommate, who had sent me his idea of a "How are you doing?" type of letter. Evidently, my ex-roommate had kept something behind since I did not pay my part of the cable bill. Hell. It was his television, and I never even watched the television, as I had one in my own room. In my own way, I made his life miserable. Knowing his schedule, I dropped by the apartment and gave a check to the new roommate – and then put a stop payment on it just to aggravate him. Call it petty larceny if you want, but there is a little saying that says "You do dirt, you get dirt." This was my way of saying in the nicest possible way, "You are the lowest form of life that exists on earth, and I don't at all feel sorry for you. Have a long, miserable life."

What I had found out earlier, was that he had a history of moving around a lot. He even admitted to me that he walked into someone's apartment because their music was too loud, and proceeded to turn their music down. Not too long after that, he must have been asked to leave, or was evicted. I don't know the entire story. What I do know, is that he had to be dominant. When he came over to my apartment to drop off what it was that he had been keeping from me, I followed him from room to room, to make sure he didn't try anything sly.

Later on, when I had my driver's license, there were inevitably going to be times that I would bump into him. When I was behind the wheel of my car, sitting at a red light, I looked in my rearview mirror and noticed a person with a motorcycle helmet. This ex-roommate used to live with my brother and my brother's friend a year or so earlier and I specifically remembered that he said that he was fixing a motorcycle. I took a closer look at this unfamiliar face and he tried to shield his face. Luckily, I was not going directly to my apartment at the time. Lucky for him, he was not dumb enough to follow me, because if he had followed me, I would have grabbed my phone and dialed 911.

One time during the time that I graduated with my two-year degree in accounting, I heard somebody yell my name as I took my seat in the auditorium during the graduation commencement. Looking up into the stands, although I can't be certain, I saw somebody there familiar to this person, who was there with this girlfriend – yes, the one who was making my life a living hell. These situations seem very situational, but I knew his mannerisms a little too well, especially watching his body language. Why would somebody who yells my name not want me to see his face? This was a red flag. I was glad that I was going to be out of the state in another month. Permanently.

Well, if I thought that that was odd, give that about three more years when everybody and their pet poodle knows who I am. That pales in comparison.

A NEW SET OF WINGS

During the time I was going to school, I took a job working as a billing clerk working at a hospital second shift during the time that I was getting my associate degree in accounting. One day, I had a seizure on the job and someone brought in a wheelchair. At the time, I was awake, and insisted that I did not need to go down to the emergency room. These people did not know me. What I thought was going to happen was that I was going to go home and get a call and be told not to return to work the following day.

That was the typical response that I got after a seizure. A lot of employers associated my seizures with the equivalent of being hung over or on illegal drugs. I fit neither of these categories.

Instead, I found the exact opposite. I found that there were people that cared. They could see that I needed help. Somehow, they had been exposed to these types of situations enough to know that there are some people who need extra help. One person that I specifically remember, Chris Peters, showed a lot of concern for my welfare.

When I was finally able to be seizure free long enough to be able to hold a license, my mom drove me to the department of motor vehicles so I could take my driver's test, which I was confident that I was going to pass. I passed on the second time – with just enough points against me so I wouldn't have to retake the test again. My mom was reading this look in my eyes and did not like what she was figuring out. What she was figuring out was that my plans were going to radically change as a result of getting my license. She was right. I had plans over the next year of slowly being able to redefine myself from the rest of my family.

Getting off the bus the following day, I walked onto the lot of a Toyota dealership and expressed interest in financing a car – that day. I had enough to put down since I had just gotten enough of a check back on my taxes to be able to afford a down payment. At this time, I had a steady job and had been working there long enough so that I knew that I would have a job to go to every day.

When I drove to my parents' house with my car, nothing prepared me for what I was about to find out. I walk in the door, and my mom runs into the kitchen, asking me how I got to the house. "Duh!" I thought that she should know how I got to the house. At this point, the only person that I told about my car, was my brother, and I swore him to absolute secrecy. And I knew that he was not a person who would go and spill the beans on me since there would have been no payoff for doing something like that.

I was probably closer to my brother than I was to my own parents, since we hung out every weekend at my apartment when I lived on my own. What was never told to me was that my mother was a closet psychic who would use that as her weapon of choice to keep me from harming myself. For those who think I am dumb enough to fall for the old catch phrase, "we sort of thought you would do that" in response to my idea to moving to Arizona, that's bullshit. My parents want to make themselves appear more powerful than they really are by saying that they know what I might do in a given circumstance.

My brother was watching the football game in the family room. Waltzing into the room, I proceed to ask my brother what was going on, as this entire situation was disturbing me. It was as though I had somebody watching my every move. Thinking that somebody is following you is bad enough. The use of psychological manipulation that my mother used was downright scary. And this is the primary weapon of choice that she uses to keep me from spreading my wings. I had a long and healthy fight ahead of me as I made it clear that I was going to become the black sheep of the family.

Aside from my cousin, who resides in the state of Hawaii since they are married to someone who is in the Army, I am the only one in my family that lives in a state other than the one that I grew up in. Everybody in my family has, for the most part stayed in one spot their entire life, living an uneventful life going to the same job every day, living in the same city, and for the most part, not learning new things. I decided that this was not for me. I wanted to get out and experience the world and contribute to it in a way that nobody else could. It was not going to be an easy ride getting my parents to listen to me.

My brother told me that mom had been interrogating him as to whether or not I had gotten a car. He said that he didn't know – I told him to say that. What was more disturbing, is that it was the mere fact that I had this car which was going to drive a wedge between myself and the rest of my family as I had to find a way to stop the breakthrough seizures from happening. That was not really what bothered me. What bothered me was how my mother got all hot and bothered by the fact that she had the suspicion.

THE CONTINENTAL RIFT

It would be this same car that would eventually give me more than I had originally bargained for. I wanted recognition for something growing up, as I didn't have as many friends as I thought I should have. I would be granted my wish, but I did not realize that I was paying a price for this recognition. It was around this time that I learned to be damn careful what I wished for. The chain of events that would follow would result in me getting that recognition and have more than a handful of people disliking me.

Was I going insane for putting my entire life on hold chasing a dream? That is unclear. What is clear, is that being diagnosed with severe osteoporosis at twenty-six years of age was enough to wake some of the doctors up. Common sense should dictate that things were going to be different from this day forward. The doctors, who spend all this time in medical school, would find that the ones that hold the key to knowledge were the patients who would shed light on things they needed to know about.

FULL SCALE WAR ON THE HOMEFRONT

I was trading my sense of privacy for recognition. What was about to give me recognition would result in people who I did not know knowing who I was. What is scary, is I don't know who these people are. Looking at this from a different standpoint, it would be this very force of power that would keep me from caving into pressure from my parents to quit my campaign after I almost died six months after moving to Arizona. Ironically, it was not strangers who I was necessarily afraid of, but my own family, particularly my mother and father, who happened to be doing their best to be resistant to change, although they have gotten used to the idea that I am embracing my new life as a freelance writer – all while not making a red penny doing it. Nobody in my family is self-employed, and I was stranded at sea. I was embracing change as I saw it happening. This made me and my parents clash like oil and water – we simply didn't have the same goals at the time. When I stopped calling home regularly, and stopped checking my email on a regular basis things changed with me. It certainly didn't include anybody in my family back in Wisconsin. Maybe if my parents understood what I was going through, that would make me want to call home more often.

I don't know who is fooling whom. What I do know, is that my potential to make a seven figure salary as a freelance scriptwriter and consultant for Hollywood had me fighting this force with all the energy that I had. At a certain point, I simply had no choice – and felt a moral obligation to find employment through non-traditional means.

Although my mom was paying my rent each month since I was not yet "established", I knew in my heart of hearts that that was going to change very soon. It would occasionally put me into a battle-of the wills with the very same people who would be looking after my care. The more I hear about vaccines for every little thing, the more the anger starts to ignite inside of me. There happened to be things that I was picking up from a personal standpoint that I could not get doctors to understand. Although beyond the scope of this book, holistic medicine was the thing that would eventually drive a wedge between myself and some of the doctors responsible for my care. I wanted to have doctors who I could trust.

Unfortunately, even if I could trust them, I knew that something was not being told to me. It all starts when I am vaccinated as a child. By the way, do doctors ever explain the side effects from vaccinations? I don't think they do – there is too much money to be made with vaccinations. And don't even try to explain that logic to my parents – they are too stubborn to believe that. I know that when they see this in print, they will just go ahead and deny that as well. They can go ahead and do that all they want. We both know who has the correct information. Sometimes being smug is the modus operandi. Get the message?

Looking for a reason for why all the chaos was happening to me, I was lucky to have a friend who was doing research on holistic medicine – who at the time met someone who was opposed to vaccination. It was her research that put me into her network – and built a foundation that would benefit me in more ways than one. Serving as not only a personal friend and neighbor that I met when I was growing up in Wisconsin, but also as a personal foundation, her experience would teach me a few things that my parents decided not to tell me about until the time of reckoning came about several years later.

SEASONAL CHANGES

The little information that my parents could give me involved them noticing that I changed anytime a weather front came through. I couldn't understand why it was that my emotional state changed with a weather front coming through. Sometimes, I would feel the fluid moving around in my head and would hold my hand to my head. There were times that when the weather would change, that I could feel the fluids in my brain moving around with the force of the Pacific Ocean in monsoon season. Anytime there was a change in the weather that brought about a thunderstorm, that was my cue to brace myself for the fall. Often, what resulted was a seizure. Every time the weather would change, my emotional state would shift. I did not know why this was happening. Doctors would not provide any type of reasoning for why it happened. I would typically tell my doctor when he asked me if I had any breakthrough seizures, whether or not I needed to count the seizures that occur as a result of weather changes. He said that any seizures should be counted. After I had my driver's license got cancelled eight months after having it, I realized that I was not in the right environment and made it clear that I was not sticking around in Wisconsin, where the weather fronts seemed to dump all the rain that accumulated over the period of a month.

As I talked to other people about why they moved to Arizona from where ever their homeland is, I hear similar stories. There are people who say that changes in the barometric pressure would cause them to have severe migraine headaches. After dealing with my own experiences, I knew that this was something that left me no choice except to move out of Wisconsin. Arizona is just like the second melting pot of the United States. There are way too many weather fronts that seem to gravitate to the part of the country where I am from. In Arizona, the skyline is blue for as far as the eye can see. Every day, it is like I am living in a blue dream. The only time that there are any cloudy days are during monsoon season, when moisture is brought up from the Gulf of Mexico and the Pacific Ocean.

FADE TO BLACK

One night, while still living in Wisconsin, I drove to a movie theatre to see a movie, just to escape from reality. Even though I was feeling a little weird that day, I went against my better judgment. When the move screen fades to black, I was not in the movie theatre – but the hospital – after going into status epilepticus. If one thing was clear to me by now, it was the fact that my difference was making me stand out – in a way that I did not appreciate. Distrustful of people, and refusing to accept help, I looked out for things myself. People who had been around that day, particularly the managers of the theater, remembered who I was because of the unusual incidents that night. I knew that I had allies in certain places – but was taking a risk at my own expense – if the wrong person saw me have a seizure. What if they did not understand epilepsy and thought that I was on drugs?

Fully aware of what I was dealing with after having my license suspended for four months, I made a decision that was going to blow open boundaries and potentially save the lives of millions of Americans. I made the decision that I was going to relocate to Arizona. My next door neighbor had relocated down there five years earlier. From time to time, she would be on my mind. Would she look any different than she looked when I knew her as my neighbor when she lived next door to me? This, in addition, to the power of the sun, was forcing me to make an executive decision. I was going to pack my bags and go on a vacation to the southwestern part of the United States. How I would get to where I would end up would be a mystery. However, it would be a mystery that would end up benefiting millions of people who were doing harm to themselves without knowing that harm was being done. To prove myself, I was going to have to go through a long period of trial and error, as I settled into what was going to be turned into a crusade to get some change implemented.

ADDICTED

Another common belief is that someone who has a seizure probably overdosed on illegal drugs. Stereotypes die slowly – others just never seem to die. The common belief is that someone who goes into a seizure – and I only know of what a seizure looks like through observation since I am unconscious when the seizure happens – can swallow their tongue. Common thinking says to put something in their mouth. I don't know who comes up with these draconian ideas, but they are from many years ago, when epilepsy was not understood. In this book, you are going to be finding that epilepsy is still misunderstood – and it still is misunderstood even by myself – as it eventually would result in me finding out through a hip fracture that I had been living with dangerously low bone density

In the beginning of time, it was thought that those with epilepsy were possessed by the devil and were colonized, separated from the rest of the population. Now, although things have changed, epilepsy is still misunderstood. All these synthetic remedies that were prescribed for people with epilepsy served to do nothing but drive the disease deeper within the body. What was never told to anybody, was that by doing this, it would result in creating even more misunderstanding among the medical community. Ever since the 1960s, when these medications were used, the only tests that were done with regard to bone density were done on those that were institutionalized, because at this point, the medications that were available did not allow a lot of people to live a normal quality of life despite their epilepsy. As the medications improved, and people started to live within the population, the issue of osteoporosis made some people take a hard look at what these drugs really do over the long run.

The fact that I had a hip fracture at twenty-six years of age – and would put me in the unfair position of fighting the same organization that was trying to protect me – the American Medical Association – after I told my internist what I believed my epilepsy to be from. It was something that also got a full-scale war brewing on the homefront, and I had to fight this war on two fronts. Not only did my parents not believe that vaccinations were connected with disease, but the entire medical establishment had succeeded a century ago in determining that those who practice homeopathy are considered quacks and not real doctors because of the fact that holistic medicine does not follow any particular science. The same people who were claiming that they did not believe that my epilepsy could be from vaccinations were also noticing that my appetite seemed much improved.

The reason that my appetite improved, was because the holistic remedies that I had started to take acted against the drug toxins in my system, which made it easier for me to stomach certain foods.

DRUGGED

I can't count the number of medications that I took to control my seizures. In a lot of cases, my appetite was affected by taking these medications. Let's face it – almost all doctors seem to know absolutely nothing when it comes to a lot of medical disorders involving neurology. The reason for this, is because different people respond differently to the same medications, whereas holistic medicine is meant to work with the DNA of the individual without causing unnecessary side effects.

I was watching the Discovery Channel and they were talking about osteogenesis imperfecta, which is a situation where those who are affected incur so many fractures that eventually they can only get around in a wheelchair. The one thing that caught my attention, was that they claimed "to be looking for the cause of the disease." This is a common theme that I hear about just about all diseases – everything from the cause of autism to epilepsy to osteoporosis. The common thread in all of these is related to the vaccinations administered to children around the time that they are one to two years old.

All the doctors know is what medication to prescribe for a specific person. Sometimes, the wrong combination of anticonvulsant medication is enough to cause a seizure to happen by itself. A lot of misunderstanding as it relates to epilepsy results because of a lack of information, even from the medical community. Doctors are just as confused as patients about the cause of epilepsy. I have the answer – and it is an answer that was not too popular with my internist who treated me for my injury in regard to my hip fracture.

All my doctors would tell me was to just be sure to get enough sleep. This is true – that stress is the number one cause for seizures. It seemed that all my doctors wanted to do is keep prescribing medication for me. They would always be putting me on different medications, under the belief that doing that would "cure" me of epilepsy. Anyone who has epilepsy, has probably been treated like a human guinea pig by the medical establishment, who have to push the new medications that continue to be released. Medicine does not cure disease – it only eliminates the *symptoms* associated with the disease, pushing them deeper within the body. In order to cure epilepsy, I would have to find a way to get to the root of the *cause.* All these synthetic remedies do is drive the disease deeper within the body, not to mention doing long term damage to the liver and kidneys.

I am going to provide an example of why medications can interact with one another in a negative way. I was already taking a high dosage of Neurontin and Felbatol, medications which are used for those with grand mal epilepsy. My seizures at this point were already under decent control, and my neurologist wanted to continue to add on more medications, with the impression that by doing so, it would completely eliminate my seizures. My neurologist recommended that I add Lamictal, an add-on medication, into the entire mix to prevent any further seizures. As I remember, he did not even warn me of the side effects from adding this medication.   That is when my brain becomes disconnected from the rest of my body. The result of what was going to happen next would make me realize that medications are not friendly to the individual.

ASLEEP AT THE HELM

Something that one of my doctors was doing, even though I was already taking several medications, was to prescribe additional medication. I was started on Lamictal, which was a medication that interfered with my ability to do things in a normal way. I drove to work feeling okay. Then the shit hit the fan. Subconsciously, I was doing things that were just out of the ordinary. My job involved a routine set of tasks, which I had committed to memory. Every day, my supervisor would bring in three to four overflowing bins of medical requisitions that I had to sort with patient insurance information which would later be entered into the computer, where the diagnosis code would determine how to appropriately bill the customer.

I was waiting for my supervisor to bring in the documents for me to sort. Out of nowhere, from this point on, everything becomes a daze. Whatever was going on, my memory was disintegrating right before my eyes. Not only could I not do my job, I was stranded. I had no way to relay the necessary information that would save me from this mess.

I had my routine down just like a juggler trying to keep all the balls in the air without dropping one on the ground. As much as I was struggling against the power of this medication, I could not do these tasks that were part of my routine. One of the balls that I was trying to keep up in the air dropped to the ground and that is when I start to go into a daze. This would be a daze that was so bad, that there was simply no way that I could perform the job that I was required to do. If this happened at a place other than where I had been working for about a year, I would have been terminated – because there is a high degree of intolerance as it relates to seizure disorders – or a tolerance of any kind where it appears that somebody is under the influence of illegal drugs. The misunderstanding of epilepsy has not only confused doctors, but also the general public, who don't know if the person having a seizure is on illegal drugs, or has epilepsy. If epilepsy is the most misunderstood disease even when we are in modern times, then the only reason to have a doctor is for the purpose of being able to legally obtain drugs.

Somehow, I also managed to eat my lunch, and throw out that, with a pair of sunglasses, as if I was sleepwalking. I don't even remember whether or not I ate my lunch. Perhaps I threw that out with the sunglasses. My battery was dead. Everything seemed completely surreal and dreamlike for me. At the time, I was working a second shift job as a billing clerk for Aurora Healthcare, sorting insurance information and patient data so that other people in the billing department could enter that information in the computer. This was scaring the hell out of me. At one point, one of my co-workers came into the area where I was working, and saw that something was not right. It was completely unusual for me to be sleeping at this time of the day. Everything in my life was in a daze, which I had never experienced before.

As I stared at the billing information that I was supposed to sort, as much as I tried, there was no way I could remember how to sort this. The result of this medication was a chemical imbalance in the brain, which affected my memory. My memory was starting to fail me. As much as I wanted to do my job, it simply was not going to happen – and I needed the money to pay my bills. What I needed was for my brain to snap out of this. It was not going to be happening anytime soon. Even if I were to take a day off, I would not be compensated for sick time since I was not an employee of the company that I worked at. My supervisor made a call to my father, who came and picked me up.

I pull out my cell phone and open the cradle to dial a phone number. Staring at the phone, I could not dial out.. As the minutes turned into hours, I realized that there was something even scarier than the fact that I could not do my job. I could not remember how to dial out on my cellular phone – worse yet, I could not manage to remember my own phone number, a number I had dialed countless times in the past. The worst was yet to come. Since my parent's number was unlisted, someone came into the room and asked me what my father's number was. I stared at them like a complete idiot. Even if I wanted to tell them, I did not have the capacity to do it. It is like I know what is wrong, but there is nothing I can do to fix the problem. With no way to give them the required information, all I can do is sit and wait for the drug to completely be out of my system.

When someone asks me what my dad's number is, I look at them, unable to respond in any way to their questions. I pull out my cell phone, doing anything to try and possibly jog my memory. I couldn't even remember how to dial a number on my cellular phone, something I had done countless times in the past. The one thing that I depended on for emergencies, I could not even use, because of a simple medication interaction. I had literally, in the course of a few hours, turned into a human vegetable. If something were to happen to me, there was absolutely no way I could communicate the vital information necessary to save my own life.

Make a note of this example. I never thought of this, but that is exactly the reason why anybody who has epilepsy should be wearing a medical alert bracelet so if you can't respond for some reason, there is a central number for people to call to get the necessary information. I can't even count the number of times that I had taken my memory for granted like it would be there for me.

Take this incident into consideration next time your doctor suggests trying out new medication. It is almost like they are using people as a human guinea pig for the American Moneymaking Administration – and I absolutely despised it. What I know, is that there is an abundance of seizure medications on the market. Why do they need to continue to mess with a system that already works just fine? Just find one or two medications that work fine with the individual and stick with it. That is not enough for some doctors. Pharmaceutical companies, which are constantly looking for ways to be greedy, moneymaking pigs, are looking for test markets for their medication. Those test markets happen to be people like you and I and countless other people with epilepsy or some other chronic condition – all without consent. It is sort of like when you were at a party and one of your friends says "Go ahead and try some of this. It will give you a high that you have never experienced before."

Finally, after what seemed like hours, my father came down to where my department was located and drove me back to my apartment and I slept off the effects of the medication that I just took a few hours ago. I slept for the next several hours. The following day, I phone my neurologist and tell him that I have no choice but to withdraw from taking this medication, after the erratic things that I noticed with my medication and how it took over my entire body, similar to the way a virus attacks an entire computer system, rendering it useless. Because I was withdrawing from a medication that my body had built up a tolerance to, I had a seizure and I am pressured again to visit the emergency room, which I had no intention of doing unless I was pronounced dead. All I wanted to do was get back home to my apartment. I didn't know what was happening with my body, but whatever was getting control of it, was causing weird things to happen. The environment that I was living in didn't make life much easier on me, either.

I was going bonkers trying to get through to the medical profession what it was that they were missing and misunderstanding. A lot of medicine in this country is not geared toward the individual. One medicine is given to every person with the same symptoms. It is like every person in the world is an experimental guinea pig. This group of people could show many different symptoms from the same medication. If you don't believe me, next time you take prescription medication, look at the number of side effects that are possible. They are about as long as a criminal's rap sheet. Later on, I would be forced to confront the whole system of American medicine, and challenge it – around the time that my life almost ended after I was lying paralyzed on my kitchen floor.

The American Medical Association was doing things wrong from day one. If anything can speak the truth about my story, it would also be from my next door neighbor, Lynn Wiese, the author of *Holistic Parenting* whose research would eventually get me involved with holistic medicine. This is somebody who I wish I had more information about. Finally, the truth gets uncovered. At least somebody has the courage to get a war started. Now, just turn over the controls and let me finish that war. I can guarantee you that I will do a fine job – and get some personal enjoyment out of it.

YES, I AM DIFFERENT

I was home one time for a high school reunion, and had a great time. I had gotten the chance to connect with people who I had not seen in about ten years. At this party, I was able to meet someone who expressed an interest in reading a couple of chapters of my book and getting it into the hands of a literary agent and editor – one that was not going to charge me an arm and a leg to get my work out. For once I was in the hands of people that cared about me and the success of my business. After all the reunion events were over, I had another five days to spend in the company of my parents house and realized that if I were going to find out anything, it was that my own mother did not understand me anymore.

She could not understand how bringing up the fact that a credit counseling service had a class action lawsuit against them was going to result in my stomach doing somersaults. Telling her that I was sick, I excused myself from the kitchen because of the inappropriate timing of the information. As far as I was concerned, it was irrelevant information and I went as far as to accuse her as finding things that were wrong with society, the same way in which she tried to find things that could go wrong when I took my road trip out to California about five months previously. As I am sitting here typing this paragraph I am keeping an ear out for if she tries to walk up behind me while I am listening to my heavy metal rock music.

Thank God she finally found something to do. It seems there is something that she cares more about besides being an absolute pest. She actually a sister in South Dakota that she has been communicating with a lot lately. At least that gives me time to write without having to cover up the computer screen every thirty seconds as she bounces from room to room wondering where to locate the address to my internist in Arizona.

She said that a scathing email that I sent to her from my laptop computer up in my room "baffled" her. As she claimed that I had changed, my knee-jerk reaction was, "Duh!". If I changed that much, then that also meant that she changed just as much. My mother and I were doing a one hundred eighty degree separation. If she was not going to understand what it was that she was possibly doing that was pissing me off, I was in absolutely no mood to tell her. In negotiation, the first rule to remember, is that the first person to speak loses. Once the floodgates are opened, there is nothing to hold back the flood of words that would have eventually poured out of my mouth.

In denial and upset about my reaction to some of the things that were going on, she finally accepted the responsibility for what had happened. I was starting to become annoyed at the fact that every two seconds, I was constantly being strung in twenty-seven directions by my mother. All I had to do was be up in my room for too long of a period of time and she would inevitably come up the stairs looking for me to see if I had for one reason or another disappeared into thin air. I had to restrain myself to keep things under control, until midway through my stay, I just blew up at her. She had no way to understand that she was putting me at risk for having a seizure from dealing with personal stress. Believe me, it has happened in other situations. I was at my wits end and was only looking out for my general health.

If she could have timed what she had said with any worse timing, then the bomb squad may have had to be called into the room. I just wish that she would understand that she is really crossing a line when she does this to me.

I tell her to just send that kind of information in an email or over the phone. She claims that I never even answer my phone any longer. When your mother has nothing better to do than hound you on the weekend, you start to avoid answering the phone.

In all honesty, before I came back to Milwaukee for this one week stay in August, I was already stressed, thinking that I was going to have more information about my screenplay and whether or not I would be receiving an option payment for it. Since two months were going by and it was still under review by the studio, I was growing frustrated more and more by the minute. I thought if another minute passes and I don't have an email from somebody telling me what was going on, I was going to explode. What good is that attorney that I retained for the purpose of negotiations, anyways? I was feeling like I was separated from society and the people who I really needed to be in touch with. I need to be living in Los Angeles and going to meetings to talk with producers so that I can sell my projects. What turned out, was that this was only the beginning of my battle. Why? The studio sent me a pass letter through email saying that my work was not marketable.

UNEXPECTED POPULARITY

When I started on the path toward getting my screenplay produced, and having my name listed with an executive producer credit all over the country as well as other foreign countries, that was the start of a rush that I had within me. I never expected that I was going to have a lot of people knowing who I was, never mind the fact that I had my face plastered on my internet website. I was getting a hint that I was getting some unusual attention from people that I didn't know – especially when they came relatively close to my face to get a look at who I was – meanwhile, I had my fist clenched behind my back in case that this person thought they were going to get the bright idea to start a fistfight with me. I really was thinking that I was going to end up getting hurt in a few instances.

By nature, I was not somebody who was a fighter, but I was prepared to go to jail and have an arrest record in defense of my life. Luckily this did not happen, since I stared this person down real well to keep a fight from starting. By the way, you can have a criminal record and still work in Hollywood writing movies. I told my father that I would have stopped at nothing to use self-defense if punches were going to get thrown. Think about who wrote all those *Godfather* movies that you probably watched many years ago.

Just for the record, I don't have a criminal record and would prefer to keep it that way. But if it has to be that way, it will be that way. While I am safely eating my sub sandwich, out of my peripheral vision, I catch somebody moving very suspiciously toward me as if they are going to go for the throat. Noticing this, I mentally prepare to very slowly stand my ground.

As I wait for the person to approach me, I watch their actions. I am right on my assumptions. I am now forced to do something that I have never been forced to do in my entire life – take a defensive posture against someone who might attack me. Just like a cat that is on the defensive, I had to make myself appear bigger than I really was and stare the enemy straight into the face with a pair of eyes that were as big as dinner plates. There was probably less than an inch between my face and their face.

Similar to a saber tooth tiger standing their ground in the front of an attack, I was not preparing to back down and was giving this person the choice to either move on or trade fists. With my eyes as big as dinnerplates at this time, I was giving the message that "I know who you are. Do this to me again, and I will remember your face." Needless to say, they moved on and did what they were supposed to do, which was order their lunch. The worst thing that someone wants to have happen is that the intended victim remembers their face. This would not be the first time.

The response I get from my parents, especially my mother, who has never been exposed to anything much more dangerous than a major traffic jam was that of disbelief. I feel that I have noticed changes in the way people have been approaching me. She is not there – I tell her about the vibe that I had one time as I walked into a theatre – which I will no longer do unless I have two people accompanying me for the exact reasons mentioned above – and she says something along the line of "Well, maybe that person was looking to meet somebody for lunch and you looked similar to that person."

Thanks, mom. When I am kidnapped, I will just have the prisoner send you the ransom note and you can feel free to pay the amount he is demanding from my hide. What she does not understand, is the same thing that I am experiencing, is also something she should be concerned about. If somebody could hurt or kidnap me, anybody in my immediate family is also vulnerable to attack. When I bring stuff like this up, it is for their own benefit as well as mine.

The amount of incredibility that I got from my mom and dad with regard to these situations was only making me go crazy. She gave a possibility of what it might have been from. She had an answer to every one of my complaints – until I made the declaration that I will call her when I am kidnapped. By the way, it is not in the interest of your health to try something like that. Let's just leave that to the fertile grounds of your imagination.

Never before in my life have I ever thought that I would be elevated to the celebrity level. Coming from a middle class family, it was difficult enough to try and convince some people in my family who have never been outside of a major traffic jam to have some empathy for someone who is trying to make a difference in the world – and finding out in the process of doing this, that he has people who he does not even know, who know him on a first-name basis. It seemed like I would be on my own, for the time being. At the time when all this shit was hitting the fan, I was going to school. I was anticipating if it would be necessary for me to drop out of school for safety reasons. I didn't know how much more of this I was going to put up with.

Some well-meaning people were spreading the word that I was a screenwriter who was writing off my life experiences after I was hospitalized for a major problem which involved having surgery to pound titanium pins through my hip.

It was during the time that I was sitting at a red light driving home from school late at night, when I noticed a person standing on the curb next to where I was waiting to proceed on to the expressway. When she started to jump up and down in front of my car, screaming like a damned baboon, I hit the security locks on my door. What I need is to have a device that shoots mace in the general direction of people that act like this. These random threats were getting me to consider one thing. There was a reason for why this was happening. I could not get inside the heads of these people and find out why they were acting this way. I figured that this was only meant to terrorize me.

This too, fell on deaf ears by my mother. "Teenagers do that a lot around here, too" is what she told me. Maybe they are trying to just put a scare into me and get some personal satisfaction from it. Honestly, there is only so much that a person can take. People that are thrown into situations that they have never had to deal with before, act like a fish out of the water, flailing around, trying to get some breathing room. I discuss this in a little more detail in Chapter six where I talk about *"The Hidden Price of Fame."*

And I really mean it, there is a hidden cost associated with being well-known. It all has to do with no longer having the privacy that you once took for granted.

When you are different from everybody else out there, there is a price that is paid for being different. We all would like to be famous and be in the spotlight. What nobody knows is the price you pay when everybody who you don't know personally, knows you by first name. When you are in my shoes and you have no way to absorb all the attention, you start to wonder if you made the right decision. I have a message to get across to those in need.

CROSSROADS

Arizona is one of the few states in the United States that does not have a high risk plan[1] for those who are unable to get insurance through normal means. In the meantime, I was depending on my insurance, which was out-of-state, to pay my medical bills with the hopes that I would get group coverage through an employer's insurance program. That never happened. After I ended up having my insurance cancelled, because of more than $50,000 of medical bills that were paid out for titanium pins which were surgically inserted in my left femur, the main bone that allows a person to walk, I had to find a way to get my expensive prescription medication paid for, at least until I could get group health insurance coverage from somewhere.

It was around this time that I stomped my foot down and realized that there had to be a different way to approach chronic disease. I was looking for a reason why all this madness was going on. Part of it had to do with the fact that I knew that the drug companies would only cover my expensive medication if I made under the poverty level. Once I made a penny over the limit, at the time for a single individual, that amount is $16,000, I had to pay out of pocket for the entire medication. That would equate out to roughly $10,000 a year for prescription drugs, leaving very little left to pay my other necessary expenses, not to mention saving for that trip to Europe for my parents who were sacrificing in a way that is commendable. I don't know how I would have been able to go through all this without their help.

---

[1] A plan that most states have for individuals who cannot get health insurance through normal means.

A HIGH PAYING TEACHING JOB

Losing my insurance would also teach me a lot about medication that doctors and pharmaceutical companies do not disclose to those who are taking the medications. I was accustomed to side effects, such as drowsiness, rashes, and appetite problems as a result of taking medication, but bone marrow suppression? I had never heard of this before, but it is a listed side effect for Tegretol, of one of the medications that I took for fourteen years. At the time, nobody even bothered to tell me about this. It seemed like nobody really cared to tell me the whole truth about what was going on. When I was growing up, bone density tests were not tests that were seen as necessary for teenagers who were on anticonvulsant medication. It would take somebody who decided to take a seemingly harmless job to deliver telephone directories around the valley of Arizona to find out that their bone density was fifty percent below normal. In the right circumstances, I might not be here to tell my story. If it weren't for the fact that I lost my job because my employer found out that I was taking anticonvulsant medication, nobody would know the truth about the extent of damage that anticonvulsant medication does over a real long period of time.

Medication does not cure a medical disorder. All it does is suppress the *symptoms* that *cause* the problem. In my case, the seizures were a symptom of a much bigger problem. But if the symptom is related to the effect, then what would the cause of the epilepsy be from? American medicine only deals with addressing the *symptoms* that are associated with a medical condition – they do not address the *cause* of a medical problem. The medication that I took only drove the symptoms causing the seizures deeper into the body, causing more long-term problems to develop as a result. The only way to get rid of my epilepsy is to get rid of the cause. I never even had a head injury in conjunction with my epilepsy.

AN EPIPHANY

For some reason, prescriptions for Ritalin, a drug for hyperactivity, seemed to be increasing. I read reports in the newspaper and on television that more and more teachers were demanding that hyperactive children be put on Ritalin. Relating this to my own situation, I knew that there was a common link to this madness. If some idiots would quit vaccinating young children, they wouldn't be acting the way they do. It was around this point, that my neighbor and very close friend, Lynn Wiese, was writing *Holistic Parenting*, a book that talks about alternative approaches to conventional medicine. All I could do is watch and learn. Her book has an entire chapter that is devoted to the damage done by vaccinations. It has a lot of people cringing. I looked on my parents' bookshelf, and they have two copies of her book – in brand new condition. That told me that they must not value her opinion. And it also tells me that they are full of shit when they say that they refuse to believe that vaccinations are responsible for why I have epilepsy. I don't even have a head injury. Something caused my epilepsy. It certainly didn't show up out of the blue.

One time when I was about eight years old, I was at my grandmother's house over Thanksgiving around the time that I was in the second grade. This was going to be the time that I started asking questions – and determining that my father was lying to me by omission. My aunt, who spends a lot of her time in a group home was at my grandmother's house during the time we were visiting. All of a sudden, she went into a fit of rage – swearing at the top of her lungs, and acting real strange. In the middle of this madness, she gets up from her chair and goes into the kitchen. My brother was afraid that she was possibly going to grab a knife. My mom grabbed all of us, threw us in the protection of the car, and we left and went over to my aunt and uncle's house without giving notice. My father claimed that the mental illness is a result of something in the family, claiming it was because she was from a screwed up family.

I wasn't going to believe any of this, even at my young age. I knew that there had to be an answer out there somewhere – and I was determined to find the answer to all these problems. Around the time that I was a teenager, I was tested at MINCEP to see whether I had schizophrenia, and the report came back negative. Even before I moved to Arizona, and all sorts of disease happened to be on the increase, I realized there was an answer out there somewhere. When I noticed the numbers of children who were all of a sudden being prescribed Ritalin, it told me that vaccination must be somehow related to hyperactivity.

It also seemed that more and more young children were being diagnosed with epilepsy, absent of a head injury. What was all of this from? As I would eventually find, there is a common denominator for all chronic illnesses. The childhood vaccinations administered in the first year of a person's life, which attack the central nervous system, are the primary cause of all neurological disorders. With all the advances that this country has made from a medical standpoint, it seems that they don't want to accept the fact that the simple ritual of vaccination is the reason why all these diseases are showing up. I started to think about why doctors were not talking as much about vaccinations. It all had to do with money. Vaccinations are what keeps the people in the medical industry employed – and the rest of the people completely confused. Welcome to the brave new world. They were administering a toxic substance, which was responsible for all neurological diseases, as vaccinations attack the central nervous system on a systemic level. This toxic brew was responsible for all neurological problems from autism to cerebral palsy to epilepsy to schizophrenia to hyperactivity and a host of medical disorders that I am unable to list. Whatever you do, don't even mention the connection that vaccination may have to disease when you are at your doctor's appointment. Somebody out there is probably going to put a hit out on my life as they read this. That's okay. The joke is on you. I already started the fire.

THE LINK

The house next door to me had people moving in and out on a regular basis. It was like a revolving door looking for the right people to occupy it – and give new meaning to the life that I was going through as a person with epilepsy. The average amount of time that people stayed there was about five years. Like a revolving door, new neighbors took the place of old neighbors. When I was in high school, Lynn Wiese and Gregg Sneyd moved in when I was living with my parents. By getting to know her, I would find an answer to my questions. She worked for Northwestern Mutual Life Insurance before becoming a full-fledged freelance author because of something that struck her husband at a young age.

BLESSING IN DISGUISE

Everything seemed to be going great for them – until her husband got myasthenia gravis. Eventually, she came to the decision to move to Arizona since the warm weather would be better and more manageable for her husband compared to the cold winter weather in the northern country. Hmm, good thinking. Maybe that will pull me out of a dangerous situation so that I can start living my life without worrying about who might be tracking my whereabouts. I was getting a little bit more attention than I was used to getting. I would often be on the bus and then have a seizure, which would definitely have people talking about me. I needed to start over. Somewhere. Luckily, the place that I was somehow pointed to by the force of God, would eliminate these breakthrough seizures – and bring about new problems that I *never* expected to have. The same age as this next door neighbor was, I found that my medication which I had been taking for fourteen years had a side effect known as bone marrow suppression. The enzymes released by this medication would prevent my body from being able to build a healthy supply of vitamin D. For now, I was just fine. But then, why do I have a persistent pain in my left heel? I would periodically notice this when I would be rollerblading down by Lake Michigan. I even notice that today, even after my surgery that I have gnawing pain that seems to run up my leg. Periodically, I will experience some pain in my left heel. There would be times that the titanium pins would grind against my femur resulting in me doing leg exercises, lifting it up and down.

At this point in my life, I can say that I was good at two things: being sad, and running away. My mom would tell me that anytime she would try and help me, for example, after I would skin my knee as a child, I would take off running down the street as a way to get rid of my problems. I knew that I wanted to run away, but at that point in my life, I did not know where I wanted to run. As a result, I would end up back in the hands of my protectors.

In the spring of 2001, I decided to take a road trip for spring break. The running away at this point in my life was going to be on a much grander scale. I was at times thinking about my neighbor and how she was doing. There had to be some reason why she picked up stakes and moved clear across the country. When I was in a position to find out, I told myself that I was going to do that. Something possessed the devil in me to dig around for some answers. What I was about to find about with regard to why she moved would communicate what was never told to me by my mom or dad.

I was giving no thought to moving down to Arizona at this point. I was focused on getting finished with my education, and did not see why I should consider moving down there. A year after having my license, I did not want to be sitting around in my apartment for two weeks straight. That is when I was daydreaming about where I would want to go. With no knowledge otherwise, I was somehow directed to the state of Arizona by the forces of a higher power.

I was originally going to see the Grand Canyon and take a road trip with a friend, which was two thousand miles. I could not talk him into going, so that is where I made alternate arrangements. Actually, that is a blessing in disguise, because I feel, had he been with me on the trip, I may not have had the time to examine how I was feeling internally. I adjusted my plans and spent time with my cousin and the next door neighbor, who now lives in Tucson, Arizona, who would later introduce me to the concept of holistic medicine.

I would also get a good dose of sunlight, which raised my level of euphoria for some odd reason and turned me into the equivalent of a pressure cooker about to explode. I was so intoxicated with euphoria – and wanted to have this on a daily basis. I thought that I was feeling good because I was two thousand miles away from home, getting to do things that I normally would not be doing. Then, I drive back through cloudy weather and rain – and my euphoria turns into mass depression. I have no way of explaining it, but the weather has a lot of effects on people from a medical standpoint. I have talked to people in Arizona who used to live in the Midwest and they would complain of migraine headaches from changes in the weather.

THE POWER OF THE SUN GOD

Upon arriving in Arizona, I would be overtaken by a power that is beyond the scope of description. Every day that I was in Arizona, the sky was bright blue, almost like I was in a dream. Upon having the sun shine on me regularly for the first time in my life, I would become a changed person. The constant sunshine would tell me what I was really missing out on. I had a stable family life and stable job back home – not to mention stable insurance because of the fact that I was a resident of a state that had a high risk program. After going through this experience, I decide to do some daydreaming. Daydreaming about what it would be like to live life as an Arizonan instead of a Wisconsinite. I evaluated my choices. I recently had to give up my driver's license during the time I was living in Wisconsin because of the weather fronts, which were out of my realm of control. After having a seizure at a movie theater which got me transported to the hospital, my doctor had no choice but to report that to the Department of Motor Vehicles, saying that he did not recommend that I was able to keep my license.

I knew that if I made this choice, I would have to throw caution to the winds and sacrifice the stability that I currently had – with no guarantees that I would ever get that level of stability back. At the time, the economy was already in turmoil, but I did not know this, because I had a stable job back home. Gee, everyday seems to be so sunny – I can't get over the view outside of my neighbor's back yard of the Lemmon Mountains.

Arizona was free of all of these weather fronts that would affect every other state in the country in the form of barometric pressure changes. Arizona has over three hundred days of sunshine every year. Wisconsin, for the most part, is cloudy and has rain hitting it all the time. It was at this point that I was starting to make a closer evaluation of what I really needed to do for myself. Maybe I shouldn't renew the lease on my apartment...

What I needed was stability from a medical standpoint as well as stability from a financial standpoint. I wasn't going to find that in Wisconsin. My decision was ninety-five percent made. The other five percent – persuading my mom and dad that the decision I was making was the right decision. How do I go about describing my feelings to someone who has never been there to experience them with me? What I was going to find was that by taking myself out of a stable situation, I would be able to challenge an organization that put millions of lives at risk without disclosure. My mom says that the drug companies had no way of knowing. I know this because I personally did something that we educated people like to call research. I looked up the side effects of Tegretol and one of the copious numbers of listed side effects was bone marrow suppression as one of the *many* side effects of this medication. The drug companies *knew* that this was a danger from the beginning of time, when this medication was released around the time that I was born in 1975 and they were in denial about the entire situation, under the assumption that once the newer medications came out, the issue of osteoporosis as it relates to epilepsy would no longer be an issue. Somebody was clearly in denial about the dangers that affected those that took the medication.

With no way for the body to produce vitamin D, the sun hitting the surface of my skin was not enough for vitamin D to be produced. Similar to the way a plant goes into photosynthesis and produces sugar after the sun hits it, allowing the plant to grow bigger, the same thing happens when the sun hits our skin. The vitamin D produces calcium, which in turn, produces strong bones, making humans less resistant to fracture.

FRIENDS IN HIGH PLACES

All I needed at this point, was the right person, with the right connections in Congress to take my situation seriously. One of my dreams is to overthrow the American Medical Association and make holistic medicine a reality in all fifty states. Well, that probably won't happen, but it would be nice. With the number of uninsured Americans growing in record numbers, both the Republicans and Democrats started to take a serious look at the healthcare system in the United States. Only one person had the nerve to take my message seriously – and that person is Senator John Kerry, who I hope gets elected in the 2004 November election. With the new healthcare reform going into shape with a new president being brought into office, this proved to be a possibility.

If you think that I am nuts, go spend some time in Europe. Over there, holistic medicine is universally practiced – and it is practiced that way in some form all over the world. Holistic medicine is a friendly form of medicine that allows the body to heal. In the sense of the word, "holistic" really can be broken down to mean medication remedies that take the essence of the "whole" person. We are the only country in the world where we allow greed to get in the way of people's interests. Profits come before people. The drug companies and doctors know this on a certain level.

Over in Europe, where my holistic doctor is located, holistic medicine is universally practiced, because of the lack of side effects – and because of its cost effectiveness. The American Medical Association and also the drug companies, are afraid of holistic medicine as it would threaten the existence of a lot of jobs. As I look in the paper, hospitals, which are supposedly non-profit organizations, seem to be growing bigger and bigger all the time. If they are non-profit, why are they focused on expansion? It seems like they are playing some sort of underhanded game of avoiding the tax man. By claiming non-profit status, and by making donations to worthy causes, they don't have to pay as much in taxes. The line of demarcation is about to get drawn again. It would send me down a path that I would least expect – one that would possibly change the way medicine is used in this country as we know it.

It would all start with one person – a person who was open-minded enough to consider that there is validity in the use of holistic medicine. My dad doesn't agree with me. We'll see who is right when the blood is spilled on the floor.

Chapter 2

The Neighbor Next Door

Foundations are necessary for any type of growth. When my good friend and neighbor, Lynn Wiese, moved to Arizona in 1996, she opened the door giving me the opportunity to start the clock over again. What drove Lynn to do her research was similar to what happened when I started to stick my nose in places where I normally wouldn't put it in the wake of my insurance getting cancelled. When I lived next door to her in Wisconsin, through my mom, I found out that her husband was diagnosed with myasthenia gravis. I had never heard of that disease. Like epilepsy, I was on the other side of people who were not aware of the disease.

In the spring of 2001, when I visited Lynn, driving like a bat out of hell for two thousand miles across the entire country, I didn't know what I was getting myself into. What I would eventually find is somebody who not only understood what I was going through, but was able to give meaning to my difference, as she was struggling through something similar with her husband. This visit would be one that would result in an impulse, on-the-spot decision. As I drove from Phoenix all the way to Tucson, and saw some pretty scenery deep in the mountain regions, something just hit me.

Somehow, as I visited them longer and longer, I was somehow getting the sense that I would have no choice but to move down to Arizona. It would be a move that would not only open my world as wide as the horizon over Lake Michigan, but it would certainly save a lot of other people from going through the same trauma that I experienced just six months after living in Phoenix, Arizona. I knew that I wanted to relocate to Arizona, but how was I going to get all my stuff down here? Oh, that's right. I do have those credit cards that still have an unused balance on them...

A BROKEN PROMISE

I know that I promised that I would try to keep the talk about other people in my book to a minimum, at its best. For this chapter, you have to disregard that rule, however. Because what I am about to talk about is what happened before I made Arizona my home.

If you would have told me ten years ago when Lynn Wiese was my next door neighbor from Wisconsin, that I would be moving to Arizona and following in her shoes, I would have no idea what you were talking about. At this point, I only had one care in my life, which was to get seizure free so I could get my driver's permit. Unfortunately, the weather patterns, combined with the fact that I was still growing and stuck inside of myself, didn't make my life any easier.

When I was in high school Lynn Wiese moved in next door to my parents. I had difficulty understanding what their motives were in moving two thousand miles to the southwest. They were growing tired of the cold weather due to her husband's medical condition. When he got myasthenia gravis, both of them went to Nepal to learn more about alternative medicine. During the time, I was still going to school working my way toward an associate degree in accounting. One thing is for sure. If Lynn and Gregg had not moved to the southwest, and if Lynn had not moved in next door to me, and had not been brave enough to challenge something that was potentially evil, I would not be here to tell my story to you. This chapter, in a way, credits her with the amount of time and research that she has put into writing a book, which talks about holistic medicine.

I would probably be working in the same job that I had been working in for two years at Aurora Healthcare in Milwaukee as a second shift billing clerk. I never would have thought that there would be any kind of connection between epilepsy and osteoporosis.

After I qualified for my driver's license, I decided that there was more of the world that I wanted to see – and I was pointed to where my neighbors had moved five years earlier in the year of 1996. Two weeks of visiting in Arizona, which was supposed to be my spring break and my hiatus, something changed. I started to think about ways to pull myself out of Milwaukee. It would not be easy as I would have to fight my mother, who wanted me to stay where I was.

I knew that I wanted to go somewhere, but at this point in my life, I did not know where I was going to end up. Part of what happened was that I was growing tired of being at the same job day in and day out that I needed a break. I had been a loyal employee for at least a year and a half and my supervisor could see that I was in need of a break and gladly accommodated my request for vacation time.

A PARTNER IN CRIME

My neighbor and good friend, Lynn Wiese, was the one who made me think outside the box, along with her holistic doctor. She had just written a book titled *Holistic Parenting* which talks about alternative approaches to modern medicine, and specifically talks about vaccines and the inherent danger associated with their use. She also served as my foundation in life. When I relocated, I knew that I would not be alone. Now that I realize it, I am glad to have moved to Arizona, because she is the one person who can understand me like nobody else can. Even in the face of change, she does not try and resist it, because she had the open mind to give holistic medicine benefit of the doubt, something that she would not have considered at her age, had her husband not been afflicted with myasthenia gravis. Because she had gone through a lot of changes and adjustments in her life, such as dealing with a sudden blow that took away the lifestyle her husband had as a result of myasthenia gravis, that made it easier to understand me when nobody else would.

Lynn could understand what I was going through. As a matter of fact, she was the *only* person who understood what I was going through outside of my internist. All the other people who had been in my life had either been overprotective of me, by not allowing me to spread my wings, or had put me in panic mode. Because she moved to Arizona, this was going to open the door to allow me to drive down there. She served as my basic foundation in life even during the time that I was getting my freelance business off the ground – or at least trying to with absolutely no success at all.

Whenever I bring up that I have titanium pins through my hip, almost everybody is misinformed. Having this one person on my side was enough to counter the countless stares that I get from strangers when I am in public. What I never knew, was that I might be in a position to make my way into the history books, Hollywood, and get attention from Congress as a result of this trip. It was starting to intoxicate me to the point where I was filled with euphoria. And I knew very well that my parents absolutely hated this idea – because it would separate the "family" from the "family" and leave behind a lot of carnage in its place.

My parents were not too thrilled about me taking this trip. "Mom, give me a break. Everybody goes somewhere in their life. And if they don't they are absolute losers, because if Lynn moved to Arizona, there must be something that she saw that would help her situation out." If I had given into my mother's line of thinking and caved, I could guarantee you that I would not be the same person that I am today. My mother is the type of person who doesn't understand why anybody moves anywhere. Everybody should just stay at home and stay in a routine.

And I am in complete misunderstanding about why I should follow what I see a good majority of my family doing – spending their whole life in one place doing absolutely nothing – except finding ways to make my life a living hell. I'm glad that I only go "home" once per year.

If I thought that the unwanted attention that I was getting from complete strangers because of my condition was daunting, I had no clue whatsoever what I was in for. As one of the only people on record with epilepsy and osteoporosis, it would be at this point that I would be defecting to the other side and challenging the practices that the American Medical Association practiced which contributed to my osteoporosis. If I never got epilepsy, I would not be on the medication that caused my secondary osteoporosis. I realize that I am not going to be in good company when I say this, but keep in mind, it is not the doctors I dislike – I dislike the system that makes vaccination a requirement to be a participant in society. Vaccination is imposed at all levels. To get your children enrolled in school, you have to show evidence that they have been vaccinated. When I enrolled at Arizona State University, I had to show evidence that I was up to date on my shots. If you were not vaccinated, you could not get educated. There must be some connection that is missed there. Maybe it means that the best education some people will get is that they don't need an education to make it in this huge world. I will get more into that topic later.

AN EMOTIONAL OUTBURST

There was one day that I had an emotional outburst after getting home from my job that I had during the time that I was in high school. My emotions are affected by my seizure activity. There is research that I have read that shows that a person's emotional state of being is directly related to their health. Very often, shortly before a seizure would occur, I would grow frustrated and aggressive.

On this particular day, I was frustrated about something that somebody at my job had said, that was not in conformity with what I believed. I stomped out of the house, slamming the back door. The end result was what sounded like an explosion, after the glass window in the door broke. Now I was really pissed. Not only did I have to deal with my emotions, but I had to wonder whether or not my parents would forgive something like this.

The noise that resulted from me slamming this door would get Lynn's attention. All the chaos involved with the slamming of the door and me running who knows where, had caused Lynn to see what happened. My mom talked to Lynn about what I happened to be dealing with. Lynn said that she was able to understand that I have my "moments". It never occurred to me *why* she understood this. There were only a few times that I had talked to her As uncomfortable as this situation happened to be, I would find out down the road that I would be put in good hands. After all, it was her knowledge that resulted in the research that would, in turn, save my own ass.

A NEW AUTHOR BORN

One day, my mom said that Lynn was writing a book. This was the first person who I knew to be an author. What could she possibly be writing about? At the time that I heard this, I was in high school. The world of publishing was completely foreign to me, never the mind the fact that I spent the good amount of my time reading each day for my classes. This was one person that I kept an eye on, to see what she was working on. When she came up with different projects to write about, as well as serving as a book publicist for Russell Public Affairs, that gave me a foundation that I definitely needed. I had somebody who I could look up to for a certain level of respect. If nobody in my family was able to understand me, at least there was one person in this world who *was* able to understand me. That was all that I needed out of this chaos.

Her book was written as a result of Gregg's bout with myasthenia gravis. However, it was much more than that. The title of her book is *Holistic Parenting*, which challenges the system which creates a lot of disease as well as the journey that she took to find an alternative solution to the problem. When she started writing a book on holistic medicine, I looked up to her for bringing attention to people, including myself, that some of the additives they put in food is not very good for you – from a neurological standpoint, especially aspartame, which is contained in a lot of diet sodas, that I see countless people drinking, including my own mother who does not listen to those people who are trying to speak the truth.

The things that she wrote about in *Holistic Parenting* made me take a hard look at the way the American Medical Association was managing to screw up everybody's life. "Yeah, we will vaccinate every newborn baby with chemical preservatives that contain mercury which is on record for causing neurological disorders in infants as well as autism and other related neurological diseases." Didn't we go through this same battle with the tobacco companies, who were selling cigarettes to underage people, so they would have lifelong customers? The American Medical Association does this same thing. They administer something toddlers, who have no voice one way or the other, under the premise that they are trying to prevent disease from occurring. To understand this, you have to know why vaccines were originally used.

Vaccines were administered in the 1950s and 1960s when diseases such as polio, chicken pox, and measles could affect the population in record numbers and wipe out a huge part of the population. The reason that vaccinations were invented was to preserve the human population from going extinct. Then, somebody had the bright idea to continue vaccinating children, even long after these diseases were no longer a threat to the population. The bright idea that followed next was even more sinister.

What resulted, was a rise in other diseases such as autism, epilepsy, cerebral palsy, hyperactivity, schizophrenia and many other neurological diseases. What a convenient way to keep those in the medical and pharmaceutical profession employed. The way that doctors defend against this one, is to say, "well, we don't have any idea what it is from." Get real! Anybody with a medical degree knows why disease is on the rise. Everybody out there seems to play dumb just so they can make a buck. For the companies that make these vaccines, there is money to be made from these vaccines. If doctors don't know what a disease is from, why did they go to medical school? I just say this because I notice that as a general answer with regard to a lot of disease.

And why is it that doctors seem to play dumb? It is because of the fact that there is too much money at stake. If nobody gets sick, there is no way that pharmaceutical companies and doctors could be in business and make the profit that they are making.

PLANET MERCURY

We advise women who are pregnant not to eat tuna, or to minimize the amount of seafood, because of the high mercury content that it has, but we go ahead and administer the same thing into the newborn child after birth – in the form of vaccinations. It makes no sense at all. The one thing that is responsible for neurological disorders on a systemic level is the preservatives contained in a lot of vaccinations – not to mention a lot of the foods that we buy in the grocery store – which attack the central nervous system. When I consulted my holistic doctor after taking the first round of remedies, he said that the mercury and other preservatives in the vaccines that I was administered as a young child were what caused a diagnosis of complex partial seizures only a year later. Obviously, it would be difficult to convince my parents of this connection, the same people that obviously have not done their research on what is really contained in a lot of these vaccinations. But, if there was no evidence of a head injury on my cat scan, the process of elimination left me no choice but to assume that vaccinations were the cause of why I have epilepsy. After all, absent a head injury, what else could have caused my epilepsy?

During my high school years, I had gotten to know Lynn – but had no idea how much just knowing her would change my life – and save it. After her husband got myasthenia gravis sometime when I was in high school, she and Gregg were looking for a way to get their life back in balance and were not finding it in the American system of medicine. They took a trip to Nepal to learn more about alternative medicine and the benefits that can be derived from it use. They learned how to use spices and herbs to cure a lot of minor ailments that come about and eventually adopted a holistic lifestyle, which is something that I slowly started to follow.

If you want to learn more about the benefits that holistic medicine has for the human body, and how the American Medical Association has found a hidden pot of gold, read *Holistic Parenting,* which can also be purchased directly off a link on my personal website. That is beyond the scope of my story, but Lynn and Gregg were instrumental in my decision to move to Arizona. Although there were other forces at work as well, such as being a little too close to the mother ship.

A NEW LEADER IN THE MEDICINAL WAR

Lynn was the first person who I knew that took the plunge into self-employment. Giving up her job in the insurance industry at Northwestern Mutual Life, a company where there is plenty of room for growth to gain additional responsibilities, to become a freelance author was bold. I also knew that there was something else at play, here, for someone to give up their job to become an author, when there are numerous other ways to make a lot of money. Somehow, I looked up to her for taking this step. I do not know why, but at this point, I looked up to Lynn as a role model, who was willing to challenge conventional methods for treating illness by writing a book. At the time when I was going to school, focusing on getting my associate degree, I could not understand what her motivations were. I read her book from cover to cover, with the interest of finding out what it was she was trying to make a point about. Even though it was difficult to understand what she was talking about every now and then, you can see the amount of research that she puts into her work.

Although she did not directly communicate this to me, she was trying to get me, indirectly, to consider holistic medicine to reverse my epilepsy. Skeptical, I simply took that advice with a grain of salt. It was a risk that I considered, but did not try out – until my life got threatened by the silent killer. It was at this point when my own life was at stake that I declared a silent war against vaccinations, especially since I did not have the support of my family through any of this. With her in my corner, I knew that she would be there even if my own family was not. I only hope that somebody in my immediate family is still alive when I get the forest fire started so that I can go and laugh at them right in their face.

BEING DIFFERENT HAS COSTLY CONSEQUENCES

In a few cases, I had my life threatened. I didn't realize that I was playing with fire and going down a path that I had no business going down. It was too late, however. The research material that had been provided to me made me realize that it was my responsibility to warn people about the connection between vaccination and disease based on something that came within an inch of killing me. There are a lot of people, my parents included, who did not believe that vaccinations were the reason why I would develop epilepsy. I hope that they are still alive when I come off my seizure medications – and no longer have to worry about having seizures. I will be laughing right in their faces about all this – and all the way to the bank.

Once I succeed in lobbying Congress to consider holistic medicine and putting National Health Coverage in place for all Americans, that will be the first step toward taking a serious look at the amount of money that not only doctors have been able to profit from, but pharmaceutical companies, and also insurance companies have been able to profit from. My parents are in complete denial about all of this research. They don't want to admit that there is a different way to do things. They fear the fact that they are going to be proven wrong. If I have my way, holistic medicine will be written into law in this country like it should be. It is much healthier for the body and allows the body to heal itself. Ask Senator John Kerry what he thinks about holistic medicine and the healthcare system in general in this country. I am sure that his wife may have a few choice words for how things need to be done.

For me, this whole line of thinking got started when I found my security badge disabled after going to a urologist appointment, as well as a plethora of medical appointments during the time that I was trying to get my feet back on the ground. My daytime planner was literally filled with different medical appointments that I had to deal with in regard to worker's compensation. It didn't help when I was told to schedule my doctor's appointments outside of the time that I was working. They could not understand why I had so many doctor appointments. If I was this different, I should just be working for myself. At least somebody will be able to understand who I am.

HEALTHY ALTERNATIVES

After Lynn and Gregg got back from their trip in Nepal, they took upon a new quality of life – which was later passed down to me when I followed their same behavior. In America, we depend on too many modern conveniences in our day-to-day activities. We eat processed food, loaded down with preservatives, without even thinking about the effects that this has on our energy level. When I started to take holistic remedies, I found that my desire for sugar was decreased immensely. My desire for food that was better for me went way up. Had it not been for Lynn's research, I never would have considered holistic medicine. My appetite also got better as a result of the use of these remedies. It is interesting that the same people who admit that my appetite is better are the same people who deny the connection that vaccination has to disease.

When I was told at ten years of age, that I would have to be on anticonvulsant medication for the rest of my life, I did not understand why this was the case. The medication does not get rid of what *causes* the problem in the first place. The medication simply attacks the *symptoms* that are the *cause* of the problem. Symptoms are completely different from causes in relation disease. These are not the same. Don't confuse these two terms. A seizure is a *symptom* of epilepsy. The *cause* of epilepsy is the vaccinations administered in the first year of a person's life, which attacks the brain and central nervous system.

After my insurance got cancelled, which was paying for my expensive prescription medications every month, that is when I declared war against the system that was fueling nothing but greed. Why support something that just pulled the rug out from under your feet? With insurance companies using my epilepsy as pre-existing condition, using that as a reason to not offer insurance coverage, the only way I could get insurance coverage was through a state sponsored high risk pool, one which was not offered in the state of Arizona.

A KNIFE THROUGH THE HEART

For three months, I had spent working at a health-food store where I worked as an inventory control and accounts payable clerk. As I am sitting at my desk, I hear the human resources person ask someone who works in my area what I am taking huge amounts of medication for. I hear him say that "[he] thinks that I have seizures." I have never had a seizure at all that anybody witnessed at this job. At the time, I was about to qualify for group health coverage – until these jerks found a reason to terminate my employment by eliminating my position. It only added insult to injury when, the day that I am terminated, the insurance cards, which I can't use, arrive in my mailbox. It seems that even the group coverage that employers offer is not really all that it seems. The people in human resources go around looking for information on people. My parents were telling me to get a job so I could qualify for health insurance. Even when I was in a job, that never happened. Even employers are looking for ways to cut the costs down. This is the beginning of the madness. I am determined to get back at a system that is screwing everything up.

By telling doctors that they were not allowed to practice holistic medicine, because the American Medical Association referred to it as quackery, that not only kept everybody in business and maintained profits, but also maintained their bottom line by charging over-inflated prices for prescription medication. You only find a system like this in America, where on the one hand, rugged individualism and capitalism prevails, but on the other hand, the individual is not cared for when a medical disaster strikes. Be sure to bring this up and be real loud about it next time you go to the pharmacy to refill your medication for whatever your ailment happens to be. Doctors, pharmaceutical companies, and insurance companies are against holistic medicine from a cost perspective. For them, it spells one thing – jobs. If there are not any diseases out in the world to vaccinate against, how is anybody supposed to make any money? Now that I think about it, that could be why I am being told that my movie does not have a market for it. Just mention the word "homeopathy" or "holistic" and the entire world goes nuts thinking that you are a quack. So, I did what any person caught in my shoes would do. I started writing to Leonardo DiCaprio's agency, asking him to get behind my script and my story.

Maybe with the right connections and appropriate administration in office to make coverage for those without insurance a possibility, this movie will see the light of day like it should. I could be wrong about this, but that could be why my movie has been delayed as long as it has been. Anybody who supports President Bush, I recommend you run and hide, because when everything is said and done, you may not have a leg to stand on. If John Kerry becomes president, we will be back where we should be.

I know about this from a personal standpoint. My dad works as a social worker, and was constantly complaining about managed care and health maintenance organizations (HMO), which dictated which doctors you could see, and how often a person could bill their insurance for services every month. The insurance companies started to tell subscribers how often they could see a therapist and use their insurance. From his perspective, his income was limited by the insurance companies ability to dictate to subscribers how often they could submit claims for therapy, and how much they would cover. That is the thing about insurance companies. For anybody that accepts insurance as a form of payment, they have to accept the rate that is given to them by the insurance company and reduce their fee accordingly. Those in the medical profession don't like taking certain insurance carriers because of their fee structure, which does not allow them to make a profit. The idea is to make a profit, not lose money.

On that note, the American Medical Association is no better than the tobacco companies. By getting people signed up for life, and have to take expensive medication, it keeps a lot of people in business. The idea is, sooner or later, a person is going to find out that they have to deal with a disease that they did not even know that they had, such as osteoporosis, cancer, or some other silent killer.

I am going on the record right now, and saying that I think that this is a dishonest way to do business. You vaccinate people when they have no way to speak for themselves, under the guise of protecting them from disease, and you eventually have a customer for life. That is the same thing the tobacco companies were found guilty of doing. It is high time that some of these people start answering questions as to why it is absolutely necessary to be vaccinating children in the first place, even after most of the diseases that these vaccinations were meant to protect against have been eradicated.

For the record, I wrote to John Kerry's office in Washington and told him if elected to consider holistic medicine be viable for everybody involved. After what I have been through, the timing couldn't be better for this, for obvious reasons.

The body can heal itself from a lot of these diseases. That is the foundation of what I consider to be the link between epilepsy and osteoporosis from a medical perspective – the childhood vaccinations that attack the central nervous system. Honestly, I don't think that doctors even take the time to advise parents that there may be a better and healthier way to deal with disease than to vaccinate their children within the first year of birth – which is around the time that I had my first seizure. When you vaccinate, you create an *artificial* immunity to disease instead of a *natural* immunity. The body has a way to heal itself. When you get sick, your body is immune to that strain of disease. I know for a fact that my mom has *not* read Lynn's book from cover to cover, especially when she said that vaccinations probably protected me from a lot of other diseases. The people who get sick the most are the elderly and young children, who have not been exposed to disease.

My father, who told me that I had an overactive brain from day one. He is right. The amount of my screaming the day that I was born was enough to wake up the dead.

MY LIFE STORY

When it came time for me to write a life story of my own, I called Lynn and asked about the steps involved in trying to find an agent. Telling me write a query letter that would get attention was not a problem for me – my situation was way too original. I do a general query on the internet and type in the phrase "literary agents". Bad move. I was directed to somebody who was not really a literary agent, but was somebody who had changed the name of their agency several times over the past two years. This agent was also at the time under investigation by authorities – but did not bother to disclose this important detail to me. Otherwise, I would have stayed away from her.

I submitted a partial manuscript to this agent and two weeks later, I received a positive response back in the mail. What I don't think that Lynn told me was anything about the Association of Author's Representatives. When I showed her the letter, she said that this person was probably going to ask me for money. Lynn was right. After this agent said that all people in attendance had approved of my work, she asked me for $300 upfront. I don't think that anybody ever even looked at the work that I submitted. At this point, I didn't need an agent. What I needed was a publisher or a producer. The agent would come much later. I was only trying to follow by example.

I had read that authors are, to a certain degree, held responsible for minor expenses such as postage and other miscellaneous expenses. The major bombshell would be dropped on me in the early New Year of 2004, when I found that my agent had her phone disconnected, and I had no way of getting in contact with her. This is covered in a later chapter, but I would learn that experience is the best teacher. I only make a mistake once. Once I am suckered, that person had better find somebody else to mess with. Being dicked with in this way only made me a lot angrier than I originally was.

SUCCESS IS BASED ON WHO YOU KNOW

Lynn served as a foundation in my life after I had my foundation ripped right out from under me. When I asked for a referral to her agent, she gave that to me. As I found out later on, based on outside factors, her agent decided to pass on the material. Little did I know, later on down the road, that it would not be her agent, but her publisher that would possibly save the day for me. One day, I was typing emails like I usually do, and there was something about the name of my entertainment company that just seemed so familiar. The entertainment company that has my screenplay is affiliated with the same company that published *Holistic Parenting*.

Lynn had mentioned to me one time that this company had moved to the entertainment area for doing movies. At the time, this seemed like it was worthless information. Little did I know that it was this very fact that my screenplay and her book in the same company would make my life a lot easier when it came to getting a production contract for *Euphoria*, a drama based on my life story, when I found out that I had bone density which is fifty percent below normal. I kept this fact in mind and saw it as a window of opportunity. Since I didn't have any connections to the entertainment industry, the fact that I knew Lynn for fifteen years, and the fact that her book was affiliated with the entertainment company that I fired my screenplay to only seemed to work for me. Personal references say a lot about a person, especially when the writer is unknown.

As I would find out, it was information like this which would make my script easier to package. I asked Lynn one time, if with her permission, I could display the book on my website, since I have a page that talks about holistic medicine. She said that was not a problem and allowed me to do that. I did this, having no idea how much of a difference that could make for me in terms of making me a produced screenwriter for Hollywood in a lot quicker time than most people would get.

Since I had her book displayed on my website, I was praying that the studio executive at the entertainment company, would recognize the title and be able to get easy access to the book. I was hoping that based on the fact that I had a long-term friendship with somebody who had material associated with Keats Publishing could possibly be the ticket to getting my script in the door. Unfortunately, my story just was not strong enough for a studio to believe in. Keats Entertainment wrote back to me and said that there was no viable market that existed for my story. Maybe once my story is in print and the public demands it, somebody will change their mind about this. Where is Leonardo DiCaprio when you need him the most?

When I was submitting my script to people in California, I had no contacts who I could network with. I had friends that lived in California, but even they did not know anybody who would be a worthwhile connection for getting my script optioned. This is an example of the fact that, in Hollywood, there is something called the seven degrees of separation. That means that anyone you know, knows somebody, who knows somebody that could get you read. In Hollywood, breaking in is about who you know, not what you know. The entertainment industry is relationship based. Who you know can make a big difference in terms of getting employment from a studio.

Breaking in, particularly when you don't live in Los Angeles or the surrounding area, is difficult. New screenwriters, who don't live in the general area have to find a way to break into the industry. Hollywood is an exclusive club that looks down on outsiders, particularly if they live a long distance from California. The fact that I had moved to Arizona made my life easier when it came to getting my script sold, but from what I have read, when you are selling scripts, it is easier if you live in the general area. Since all of my doctors are located in Arizona, and at this point, I was not going to move to Hollywood with no contract on the table, I had to find a way to break in.

By the time this book is in print, and bought by a publishing house, there is a high chance I will no longer be living in Arizona. I will be living in the city known for turning dreams into reality. I tend to wear out my stomping grounds every three years anyways. That has been a general trend – because I have to find new ground to stomp around on and create the amount of ruckus similar to a pack of bulls charging down a football field.

Lynn's writing provided a necessary foundation for my success from a financial standpoint. Using her network of contacts, I was also able to get access to certain people that I wouldn't otherwise have known about. For example, her holistic doctor, who would end up teaching me about the benefits of holistic medicine, I found through knowing Lynn. The same thing goes for all of her professional contacts. Anytime I am looking for somebody to possibly add some credibility to my story or whatever I might be writing, I call Lynn. Chances are, she will be able to point me in the right direction. She served not only the role of a friend, but also as a role model that I looked toward when I was writing about my experiences.

Her success as a freelance author and the fact that she now works in the area of literary publicity for Russell Public Affairs would benefit me indirectly as I started to learn by example. There were so many questions that I had unanswered that I should have asked before I got signed an agent. Questions such as why an agent is not supposed to charge you money upfront before a sale is made.

A bad foundation results in a lot of bad fortune. When you are first buying a house, you check to see if there are any leaks in the foundation. The last thing that you want to have is a flooded basement after a rainstorm. The same thing can be applied to real life. If there is no foundation in place, a person will inevitably fall through the cracks and not be rescued. The world is a very cruel place – particularly when it comes to writers. If it were not for Lynn being a published author, I would not have the contacts that I desperately needed to forge ahead in the world.

It was also through Lynn that I got information about her holistic doctor, who is working to help cure some of the ailments that I am dealing with. When I noticed that my appetite seemed to be better than it was before I was on holistic remedies, I saw that there was value in holistic medicine. It made me wonder why we have such an inefficient method of treating those who have diseases.

It seems interesting that the same people who said that my appetite was better were the same people who denied the connection between vaccination and disease. Some people just never want to know the truth.

Chapter 3

Winds of Change

A person who lives their life in one place their whole life is dull and boring. A lot of times, my emotions were a rollercoaster ride. As I entered Arizona, somehow, I became a lot calmer and everything was on cruise control. In trying to escape from the demon, I would end up in the drunk tank twice. Both of these times gave me time to reflect on what I really should be doing besides the field that I was trained in. As I came to the reality that somehow my anticonvulsant medication was linked to secondary osteoporosis, I realized right then in rehab that my life would never be the same again.

As I drove through the Colorado Rocky Mountain area in Grand Junction, I couldn't help but take note of how pink the sky was – and the fact that there was an incredibly pretty sunset. At this point, my life was starting to hit cruise control – but I did not have any idea where I was going at the time. Little did I know, that I was very soon going to run into a dangerous situation which would leave my life hanging in the balance – and would result in a declaration of war against something that was supposed to be helping me.

After my move to Arizona, I would be taken into custody and spent time in the drunk tank several times when my seizures were out of control. It was during one of these times in the drunk tank, that would get me thinking about what my real objectives in life were supposed to be. It was nothing to be taken lightly. I was going to gain knowledge without even having a medical degree. No form of education would prepare me for what I was about to go through next. It would tell me that I was really different and that I did not need to have that sheepskin.

It was the last day of school before my spring break came up. Happily finishing up my classes, I went home and finished packing for my trip. Proceeding to drive out of town, there was not much variety. As I drove through Chicago and the Iowa, all I saw were long stretches of farmland. When I reached the Colorado Rocky Mountains, it hit me. I saw a beautiful orange sunset, which was so bright and vivid that it was slowly changing the path that I was leading in this life. The most scenic part of my two thousand mile drive was the drive that I took all the way through the Colorado Rockies.

Driving through small towns, I thought to myself, I could live in one of these towns. Then it really hit me. Hard. My emotions before I left Milwaukee were in a state of turmoil all the time and changed on a daily basis similar to the constant weather changes and fronts that came through Wisconsin. As I drove through the Colorado Rocky Mountains, I was for the first time in my life, seeing something that would give some meaning to my life. With the sun as my guide, my emotions eventually were put on cruise control instead of putting me on a rollercoaster ride going from one extreme to another extreme.

I was always attached to the environment. Before I got my driver's license, my dream would be to find a way to save up for a boat and visit some exotic place. Since I could do that without the need for a driver's license, my dream was to somehow get the money for a fancy boat and sail the Mississippi River. My father had always talked about going to Lake Itaska, which is at the mouth of the Mississippi River. I had that very thing without even leaving the country. As I drove through the mountains, the idea of living in the mountains was slowly creeping up on me. But how would I be one to make the adjustment?

I could not understand it at the time, but my euphoria that I was feeling was from something supernatural and Godlike. It was as though the euphoria was taking control of my decision making process. Well, I was on vacation, what else could it be from? When I got back home from Arizona, driving from the tropical sunlight into clouds and rain back to Milwaukee, my euphoria took a dive into mass depression again. After making this observation, I started to examine my options. My lease was coming up for renewal on my apartment. Before I left on vacation, I hastily signed the renewal – not realizing that I had a change of heart about the entire situation. What I thought was going to be a vacation to get away from everyday life turned out to be anything but that. Somehow I knew that – it was the first time in a while that I was intoxicated. After I got back, I sobered up – and got hit with mass depression.

I notified the landlord, that I had to change my mind. With only sixty days to plan, I tried to take care of the specifics – get a job, find a good living environment, a referral to a doctor. I found that I couldn't get a job as an out-of-state resident. It was difficult to get a referral to a doctor, since I couldn't get my doctor to refer outside of his network. At this point, my insurance was out-of-state through a high-risk fund for people that can't get insurance through normal means.

At that point, I told my parents, whether they liked it or not, I was planning on moving after I finished up my degree in accounting. I was taking a big risk that would open my life up in a way that I never thought to be possible.

SACRIFICE

I didn't know what I was getting myself into, however. In Milwaukee, my social life, sucked, to say the least. The only way that you could have a social life in this town was if you were a heavy drinker and liked to be a party animal. That was not for me – for various reasons. My medication made drinking prohibitive – which most people simply did not understand. To be part of the "in" crowd, you had to do what is referred to as "having fun", in other words, getting drunk. I just wanted to be somewhere that I could start over and be happy. And I would find what I wanted right in the heart of the Arizona desert. I found that with the numbers of people here, it was like a second melting pot of the United States. I could be myself without worrying about fitting in to any predefined mold. I don't need alcohol in order to feel like I am euphoric – I get that feeling when I am outdoors in the sweltering Arizona sunlight.

I would be trading my financial security for a better social life. It was as if I was stuck in a time warp. The grass is not greener on the other side of the pasture. I may have more money than I know what to do with, but again, come new problems. That is all an illusion, until you cross over to the other side and see that the grass is a lot more colorful. However, I was ready to take this risk, as it would allow me to start over from square one and re-establish myself in a new town, where nobody knew me – and it would allow me to have a better foundation to operate from.

As my car zoomed down the highway from Flagstaff into the heart of Phoenix, that was when my situation took a major turn. I could feel my head filling up with so much pressure that I thought it was going to explode. For two days straight, I suffered with a migraine, which for some reason, kept my seizure medication from working the way in which it was intended to work. Popping Ibuprofen tablets that I normally stock in my bathroom medicine cabinet, I expected that with a full night's rest, my headache would go away, and I would stop having these damn seizures. At least temporarily, it seemed that my seizure activity increased.

There was a time issue. I had been in such a hurry to get to Arizona, sign the lease, and get my keys to my apartment, that this was most likely brought on by overstressing my system. Just trying to go from Albuquerque to Phoenix was an eternity, even doing 90 miles per hour. The last thing that I wanted was to get to Arizona and have no way to move my stuff into my apartment. My brother and my father were booked on a one-way ticket via Midwest Express back to Milwaukee and for that reason, there was pressure to get moving quicker. It seemed like when I made the trip, I could do it in two days with no problem. The one thing I didn't take into account, was the time involved in unloading all my belongings. Even with a two hour difference in time, that still did not buy me enough time. The time involved in staying in hotels, regular stops to refill the car and moving truck with fuel, and pit stops to re-energize ourselves, there was simply no way to account for how much time it would really take to move everything down to my new apartment. It seemed like when I made the trip to Arizona by myself, I could easily do it in roughly two days, driving close to twelve hours straight, relying on caffeine to stay awake.

With no time to clean my apartment, my mom stayed behind and cleaned my apartment after it was empty. Although it was depressing from the standpoint of my mother, the move would open up my world in a way that I could never imagine.

After unpacking all my stuff in a disorganized fashion, driving my brother and father to the airport, my priority is finding an accounting job. I wanted to give some meaning to the education that I got in Milwaukee. At the time, I thought that I would get trained in the area of forensic accounting, and get a job that allowed me the freedom to travel and see the world via an international assignment. That may eventually happen – but not in the way that I plan for it. I would find out that with Arizona being highly transient, a lot of employers were leery of people who had degrees from out of state schools. I had a massive migraine headache that would not go away, even though I was taking more than the regularly suggested amount of headache medication. To make matters worse, I had that butterfly feeling in my stomach – and had to find out how to get to my job. Actually, when I got behind the wheel of my car that night, what I didn't know was that I actually was going to my job.

As disorganized as I was in my personal life, I was also completely confused about where my true job was going to be found. And it was not in the field that I graduated in. It would be something that would not only open my mind up to the endless possibilities involved, it would also give me a reason to get up every morning. Actually, this job would keep me up, not allowing me to get any sleep, as I was determined to make an impact similar to the black mark that was left on the planet of Jupiter. I would, for once in my life, be involved with something that would change the entire world as I know it. Who ever thought that by moving to the desert region of the United States that I would eventually set the world on fire? That could explain the reasons why there are all these random fires that pop up on the local news every so often.

Slipping behind the wheel, I drive off into the night, and end up getting lost. Unfamiliar with the way the freeways are laid out, somehow, I end up thirty miles or so from where I am supposed to be. With a case of the butterflies, I exit off the freeway before I cause a major accident. The next thing that I know, I wake up and notice that my car is sitting in a drainage ditch. I left the car in gear – as there is damage to the left side of the car. There is glass all over the front passenger seat, after my car slams into a tree. As I am in a drunken stupor, I wake up and the first thing I notice in front of me, is this weird material hanging from the steering wheel. I look to my right, and the same rubbery substance is hanging out of the glove compartment. Am I in the damn twilight zone? Finally, it occurs to me that these rubbery-foam substances are the airbags, that supposedly went off when my car landed in a ditch. Since it is dark outside, I have no way of knowing where I am relative to where I need to be. And that is where I get completely confused.

BABY STEPS

I ask the police officer why the hell my car is in the drainage ditch. A civilian, who called the police, after hearing the airbags on my car explode with the sound of gunshots, says that he saw my car slam into a tree. As I try to sit up, my back is extremely stiff. I don't want to move. Of all the seizures that I have had in my life, this seizure is the most painful that I have ever experienced in my life. As I am loaded into an ambulance, I have so much pain in my back, that I require the assistance of the paramedic to move from my car onto the stretcher. I can't even understand why these seizures are happening. It's as if my battery is dead and can't be charged. As I am taken into the custody of the paramedics, my back is hurting so much that I have to take short steps. It takes two people to assist me as I take these short baby steps. All I had was a seizure. This was unusual. Even when I would have seizures in the past, there would not be this much pain. This would be a pain that would revisit me later on when I least expected it to happen.

A NIGHT IN THE DRUNK TANK

I am kept in the hospital overnight, with IV fluid put into my system. Whatever is in the fluid must have stabilized me to the point where I could resume a normal life without worrying about drop seizures. My life becomes normal – now all I have to do is find some form of employment since my mom has been paying my rent every month because I can't pay it due to unstable finances – my mom is still paying my rent as of the printing of this book, and I have promised her that I am going to pay her back in full as soon as I get "paid".

I did not want my mother to find out about my overnight stay at the hospital, so I just made life simple and billed all my hospital bills against my car insurance, which at the time ended up being around two thousand dollars. In the midst of all this, I mention to my cousin that I had an accident. Without realizing it, the word got around to my mother in a quick amount of time. In the end, the insurance company totaled out my car, which resulted in an $8,000 claim to my insurance company – not to mention the $2,000 in medical bills that I made them ultimately responsible for.

LOST AND CONFUSED

Getting home from the hospital, I try to rest and still have a stiff back, but it goes away after two days. Thank God, at least give me an opportunity to get back on my feet. With no car, and at the time no job, I was stranded. Eventually, I used my high credit rating to be able to lease another car. I was met with resistance at one dealership, who had a problem with people who worked through temporary agencies. So I found another place that would take my business and I made myself look better than I really was at the time. I was put into survival mode before I was ready. Either get reliable transportation or fry like an egg on the sidewalk. I remember the first summer I moved there, was probably the highest on record, at least for the time that I was a resident. There were a few times that the temperature got as high as 125 degrees. In the time that I have been living here, I have watched as my skin changed from pasty white to being a dark orange color.

This was why having a pool in my apartment complex was a priority at the time. As I walked from my apartment to the pool, I came upon another realization. The sunlight was coming down with so much power that it made walking barefoot on the pavement very dangerous. It was almost like playing hot potato. The sun was going to become a central figure in turning me into a teacher. Little did I know that this would be the profession that gave me the stability that I needed. I didn't realize that teachers were compensated at such a high rate of pay. What was never told to me was that one of my medications that I was on for fifteen years had resulted in bone softening and a fifty percent below normal bone density. This was one of the conditions where I was not showing any symptoms but living with a condition that, in the right circumstances, can result in death.

TEN DAYS OF HELL

It was the beginning of September, two months after I got settled in Arizona. I started to feel pangs in my stomach and had a severe aura, with no seizure. When I moved, there was nobody that told me that the economy was in unrest. If it was at that time, the worst was yet to come. I could actually feel it happening, but did not know what was going to happen. Evidently, somebody from Arizona had faxed a concern to the FBI field offices that there was a concern that a growing number of Arabs were showing interest in flying commercial airplanes. I wish that I would have been able to relate my experiences in the beginning of the month to the people in Washington.

It was the first day of September. My personal radar was picking up something very, very weird. Somehow, I knew that September 11, 2001 was going to be a *really* bad day. Nobody had to say anything to me to let me know that this was the honest truth. I had a premonition of something terrible happening, but wished I had more information. Even though I was over two thousand miles away, it was like I was able to sense pandemonium happening. It was getting so bad that I was about to throw up. Actually, just two months earlier, I had moved to Arizona and had gotten in just under the radar of all hell breaking loose. I never mentioned this to anybody because everybody would have thought that I was nuts.

It was a Thursday, which was the day that my temporary agency had sent me on an interview to Hi-Health, a regional health food store. Two days earlier, our country was attacked by terrorists that crashed into the world trade center towers. For me, this has a lot of significance because of the event. I can guarantee you, that almost everyone remembers where they were during the time that our country erupted into chaos. If you ask anybody what they were doing or where they were during this time, I am absolutely sure that everyone would be able to tell you exactly what they were doing, or where the were at the time of the incident.

A lot of people who were not at home to witness the events unfolding on television thought that this was a repeat of the 1993 attack on these same towers when somebody let a bomb explode in the parking garage. When I turned the television on and saw smoke coming out of the world trade center towers, I almost threw up on the floor. That was pretty much the only news that was every reported for the next week straight, and that is when our country saw a new world order come to life under the current Bush administration, who was starting wars against every country, costing Americans their jobs, and he was doing this for one reason. He did not know how to run the country. By getting our country involved in so many wars, our country went to hell in a hand basket. His whole idea of maintaining job security was to make sure that our country was involved in so many wars that when the next election came around, it would be impossible for the rest of the country not to re-elect him.

I was not one of these people that was going to be fooled, because I knew that this was the real thing. As I sit down on the floor of my carpet, I start getting pangs similar to being nauseated. But, I couldn't throw up. My body was trying to expel excess energy that it was dealing with. However, this energy did not exist. My own body had been fooled into believing that it was sick by means of an aura. This had the same impact that the Kennedy assassination had on my parents' generation. Even though it has been almost forty years as of the writing of this book since John F. Kennedy was assassinated, if you ask anybody in my parents generation where they were, or what they were doing during the time John F. Kennedy was assassinated, they would be able to recount the details vividly, by telling you where they were and what they were doing during that time.

It is the same thing with me. The days surrounding September 11, 2001 were so vivid that I am able to remember what my activities were around that time. I was at home and got up as usual, turning on the television to the Fox News Channel. I couldn't imagine what I was seeing and had to verify for myself if I was really seeing things. I picked up the phone and called my parents and wanted to see if they were aware of what happened.

This was the beginning of hell. This was the same week that I was interviewing for a position as an accounts payable and inventory control clerk at the corporate offices of Hi-Health. Before I had moved, I had a stable job while going to school in Wisconsin for two years. I found that I had trouble making car payments, not to mention covering the remainder of what my car insurance didn't cover when I trashed my first car.

The other thing I remember around this time was that the following Saturday, September 15th, my father was due in on a flight from Midwest Express. Shit. With all planes grounded in the aftermath of the terrorist attacks, I was wondering whether or not he was even going to be able to come out and see me. I was more concerned for the well-being of my cousin, who is a pilot for America West Airlines. You can only imagine what was going through his mind as was grounded for three days in Albuquerque. He had no idea what was happening. Since the planes that crashed into the towers were hijacked from all over the country, I didn't even know if he was a pilot on one of the planes. With no information about what airlines were involved, the only thing I could do was pray. If he were on one of those planes, I probably would have found out, since my mom regularly talked with his mother on the phone. The only thing that he was told to do was to land at the nearest airport and call his supervisor. My father didn't even know whether or not he would be able to rebook his flight to come out to see me, since all existing reservations in all the computers had to be deleted because our country didn't know who was a terrorist and who wasn't a terrorist. Luckily, he was able to rebook his flight within a few days, after the FAA released its suspension on air travel for the three days that all air travel in the United States was suspended. The FAA had to implement a new screening process for every passenger going and coming through the airports – myself included.

This would give special meaning to the word "different", as I was concerned that the pins that were put in my hip in the aftermath of my fracture later on, would be enough to set off the security alarms at the airport every time I passed through the gates. In a couple of cases, when the security personnel waved a wand around the area where the pins were surgically implanted, alarms sounded and they asked if anything was in my pocket. For a period of time, I carried a note written by my doctor saying that I had surgical implants. Luckily, for the most part, as I was pushed through the airport by means of a wheelchair by the skycap, I didn't have to go directly through security. Since I had my crutches with me, there was no way I could manage carrying my bags if I were unable to put any weight on my foot during the healing process. The last thing that I wanted to do was to shatter the bones that held the pins in place and have to have a hip implant put in, which wears out every few years.

At least I could use the fact that I was on crutches as an excuse to bypass the line at the gate boarding the plane and get on the plane before anybody else could. Even today, when I am carrying the bag, which holds my notebook computer, and a personal bag which holds a lot of reading material, I use that as an excuse to bump ahead in line. Sometimes you have to look for ways to violate the rules. As far as I am concerned, it is perfectly legitimate.

SETTING A FIRE

What nobody told me was that I may be somebody to hit the history books – based on taking a job that I normally would never take. I get a job-in-training through a temporary agency working as an accounts payable and inventory control clerk for a health food store. I got hired as a result of two weeks of working hard for them. This would only last three months, until it was found out by those in the human resources department that I have epilepsy. I started working in the accounting department, but then worked full time in inventory control where my responsibilities were following up on confirmation orders and doing special orders for the stores. I have always been someone who likes to be in a routine and that is what I was looking for in this job. For somebody with an agenda, a routine is boring. As it would turn out, my job would be eliminated. Don't kid yourself that I don't know why it happened that way. They just didn't want somebody who was taking medication working at their company, so they find a way to fire me.

Before I moved to Arizona, I had worked for two straight years, with no problem having to worry about car payments, rent payments, or other obligations as they came due. I was hoping that I would find the same thing from this job. The only problem at the time for me, was that the only way that I could get insurance to cover my medications is if I have group health insurance. With no state high risk policy in place, it's even a question as to why we have the problems with healthcare that we have today. With only thirty-one states having a safety net of some sort for those who are unable to get any type of insurance, the ones that don't have a safety net are the ones that usually have the biggest problems with healthcare. Until we can get somebody to put national healthcare into place, these high risk pools are the only option for people who can't get insurance through normal means. I have included a list of these companies along with the contact information in the appendix at the back of this book.

The education that I would learn during the time I had this job at would be a preparation for my next job. As I got home and started to cook my dinner one night, I have no way of explaining it, but it felt like someone had just come up behind me, and shot me in the back with a gun. Doubled over and on the floor, I crawl over to a folding chair that sits by my desk. It was a high stakes poker game that was being played with bets as to whether or not I would make it through the night. After two hours of being slumped over a folding chair, pushing my hand against my lower back, I was finally able to stand, holding my hand to my back.

Whatever had just happened to me was at the time beyond the scope of my imagination. It occurs to me that my body had released calcium from the skeletal system, and without my knowledge, it was going to result in a kidney stone. I came to this conclusion after the pain miraculously disappeared and I was able to resume a standing position again. I must have had an ace up my sleeve somewhere, because I was able to pass the kidney stone and not have that result in something serious. The last thing that I wanted was to have to go and get medical treatment and get a battery of tests done when I had no idea what I was dealing with.

A SELF-REALIZATION

Finally, I get my ability to walk after two hours of dealing with a back problem. Giving myself some head time, I realize one of two things. Either I just inherited my uncle's back problems, or this was a repeat of the car accident that I had only two months ago, or none of the above. Heading to the shower, something happens that makes the hair on the back of my neck stand up.

When I went to the bathroom that night before going into the shower, I noticed something that just made my hair stand on its end. The water in the toilet was a dark red color. And I just had a painful urination. Shit. During the months ahead, I would have a painful urinary tract infection as a result of discharging a kidney stone through my urinary tract. Thinking that this was actually related to my car accident that I had only three months earlier, I disregard this as mere coincidence and went on with my day-to-day affairs as an accounts payable and inventory control clerk for a regional health food store. Maybe God was not sending a strong enough message that I was really supposed to be a teacher instead of working in an accounting job.

Since both of these situations passed without me having to go to the hospital, at the time, I was winning the game of poker. Who am I kidding? I have never even played poker in my life. God has his ways of sending a message when he needs to send a message. I was not listening – I was healthy and had no reason to be concerned about my life. If God wanted to send a message, he was going to have to get creative about the way in which he did this. The heat would eventually get turned up to a high pitch as I would have the choice of either going to the hospital or surrendering to the man upstairs.

My mother was the one who looked further into why this was happening to me. At the time, there was not a lot of information about what was causing this to happen. Her research indicated to me that people who get kidney stones, usually get more kidney stones. What her research indicated, was that the kidney stone had been from a release of calcium, which built up as a calcium deposit. I had cheated death by not having to be operated on to remove the kidney stone. My goal was to avoid an operation of any kind, since I had out-of-state insurance. My mom had told me that usually people who get kidney stones are more likely to see additional stones sometime down the road. When that happens, is unknown.

When I was in the hospital after my car accident, the hospital took x-rays. I thought that they would have been able to find something. I started to pester them about if anything was found on the x-rays. They tell me that no fractures were found on the x-rays. Whether or not this was the truth, I won't know for sure. Then, if there were no fractures, why in the name of God did I have another instance where my back just locked up on me? There had to be a reason for all this mayhem. It would turn itself up in its own little ways. What the hospital did not run was a bone density test, which would have revealed that my bones were brittle and likely to fracture in the right circumstances. This would also reveal that I would be the one who would set a landmark case, being the only person with epilepsy and osteoporosis. Since a lot of the doctors in the United States were not aware of the link that existed between epilepsy and osteoporosis, bone density tests were not regularly done on people like myself.

VISIT FROM BACK HOME

In February of 2002, my parents flew out to Arizona, since it is seasonal for me but absolutely cold up in the north country. In Arizona, the seasons are screwed up. The summer season in Arizona is the equivalent of the winter season in Wisconsin, just to the opposite polar extreme. A lot of the time in the summer, I spend inside and only go out when it is necessary, like to fetch the copious rejection letters that were filling up my mailbox as I tried to get a message out to the world. The problem was, nobody thought that I had a chance in hell of getting my message out and that I lacked any authority to get my message out. The same people that were screwing everything up for everyone who was on these medications are the same people who are considered to be any real authority. Gee, we have a nice system in our country.

GETTING TO THE TOP OF THE MOUNTAIN

The plans were to go hiking with a friend that I had met at church a month earlier. Getting to the top of this mountain was like a victory. Having a full view of the entire city and having my picture taken up on the mountain gave me a sense of power that I did not yet have. Somehow, being on top of this mountain gave me more freedom than I initially had. For the first time in my life, I was on top of a huge mountain that overlooked three million people. Not only was this exhilarating, but it was the first time that I had seen a view from up high. That was where I was going to set my sights. Up real high so that I could regain my financial security. That power would be one that would be taught to me through self-discovery. There was something giving me an indirect feeling of invincibility. It was all an illusion.

Climbing up the rocky path, the only thing that worried me was falling on my face as a result of a rocky path. A steep climb to the top would be well worth the effort. The climb up was easy. The climb down, on the other hand, was not so easy. If I did not place my feet the proper way, I would end up falling flat on my face all the way to the bottom. Since my parents, particularly my mother, is overprotective by nature, she was telling me to be careful in the case that I had a seizure in the wrong spot. I felt that I knew myself better than she did, and if I were going to have a seizure, I would know it. I had not had a seizure since my car accident. And I was not planning on having another one on my watch.

After the weekend was over, I told my parents to have a safe flight and to give me a call when they got back to Milwaukee. As far as I remember, I got a voice mail from my mother complaining about how the weather sucked in Milwaukee. It was always raining, snowing, or overcast due to the weather patterns that seemed to always spiral towards that region. Being safely tucked in the south of the country, I am able to avoid most of those weather patterns. Except for monsoon season. That is when mother nature turns every movable thing on a patio into a deadly projectile – and sometimes turns me into a moving projectile after I have a seizure due to the barometric pressure changes – but these instances nowadays are far and few between. I don't even bother to clean my patio furniture out on my balcony, because it will just get dirty again anyways. Arizona is known for transporting dirt and dust, being in the desert region of the country. I personally remember one of those dust storms because I happened to be unfortunate enough to be caught in one during the time that I was on crutches.

Eventually, I am able to get my parents to understand what it was that made me decide to jump out of the security that I held working as a billing clerk for two years and move two thousand miles away, at least for a little while before she became upset with the lack of communication. I certainly had a difficult time explaining it. How do you explain to those who have not yet been touched by an awesome power that you just simply feel better where you live? I knew it the minute that I visited Arizona. It was what directed me to this state that is even more questionable. Why in the hell was I here in the first place?

Many times after a major rainstorm, I would come outside of my apartment and the entire yard would be littered with tree branches. These are not small branches, either. It is enough to keep the maintenance people busy for a good majority of the day cleaning up the mess. When I am sitting at my computer, which faces the patio window, there are times when I am expecting rain and what do I hear? Not rain – I hear hailstones hitting my patio window. Somehow, whenever there would be a cloudy day in Arizona, I would mildly feel the effects of this – but not to a point of a migraine headache, or a seizure.

When I was living in Wisconsin, the demon would have a way of playing with me by turning the fluid in my brain into a turbulent ocean resembling the Pacific Ocean during monsoon season. Since this could not be accomplished with any degree of success, the demon just decided to lay in wait for a different situation to arise.

JOY THROUGH SUFFERING

My mom had been paying my rent every month because I was unable to find a job that would be enough to pay all my expenses. I was growing tired of this. I knew that sooner or later, her job would end up getting the axe because of the fact that she was not getting the volume of work that she normally got. With the advent of the internet, there was not as much of a need for an executive search specialist, which is a person who does work locating a doctor to work in a specific area.

When my unemployment ran out, and the extension on my unemployment was not there like it should have been, I had to depend on my parents to help me out. You can imagine the amount of times that I pointed the finger at the Bush Administration and said that maybe he should have been the one trapped in the human death trap and allowed to rot with the rest of three thousand other people. As a result of his policy of only looking after the wealthy, which at this time, I was not in any way a part of, I was not receiving any help anywhere. At the same time that I was growing frustrated with the fact that my temporary agency was not calling me back and sending me on assignments, my mother was also not getting the volume of work that she had been getting. I probably did not do the right thing, but I decided that it was futile trying to look for jobs that didn't exist.

There were a few things that I was going to find out that hurt my chances – and help them at the same time. Just to help out, I was ready to take any job that came along, even if it meant doing physical work. I took a job to deliver phone directories to homes and businesses around the valley. And that is when my priorities took a radical shift. It took this self-discovery that would not only teach me about myself, but would also allow my parents to save money – and provide for a all-expense paid trip to Europe for my parents when the time would come. They were planning on going to Norway, but when they had to pay my rent and help me out every month, it ate into their vacation budget. I told them that when I had all eight juggling balls back in the air that I was going to pay the entire cost of a trip to Norway out of my bank account – after I finished serving my sentence. Before any of this even happened, I was going to have to do something that would almost kill myself.

THE DAY THAT LIFE CHANGED FOREVER

It was a Saturday afternoon, a day when I normally would not be doing much of anything. Except for today. I was at the loading dock trying to put phone directories that weighed as much as fifteen brick cinderblocks into my car – and acting like an imbecile at the same time. It was very hot outside, with the hot Arizona sun beating down on me. Trying to lift the bundles of phone books, I found that it was impossible to lift them. As the afternoon approached, on a 100 degree day, my temper is about to flare as out of control as the heat happened to be that day, I decided to lift them differently. Using the upper part of my leg, I heaved them into the trunk of my car. Driving back home, I was very tired. Exhaling rapidly, I just wanted to get home and crash. Climbing up the steps to my second floor apartment, exhaling in and out very rapidly, something did not seem right. Was I ever this tired? I have been through similar situations and not been so tired.

This is the point where I learn more about something I was never told as a child – that one of my medications that I took as a child for fifteen years had made my bone density level dangerously low. This is the first moment when I would find out the reality of a lot of the earlier anticonvulsant medication that I had been taking as a child. I had been living with osteoporosis for fifteen years and did not know about it – until the movie screen faded to black and I collapsed on my kitchen floor. Finally, I figure out why it was that I would have pain going through my heel after a long session involving rollerblading down by the lakefront in Milwaukee.

After I got home, I was breathing and very much overtired from the physical labor. As I tried to sit down on the floor of my apartment, I was in for a shock. As I started to squat down on my floor to catch my breath, I could feel my left femur rubbing against something, as if it was out of place. I never heard a popping sound when I lifted the phone books. There were times before when I would be walking that I could feel my femur rubbing against something when I was younger.

Thinking that I simply pulled a muscle, I just took some pain medication. At a certain point, the movie screen went dark – and I got a visit from the ghost of Christmas Past. That is when I had real big problems. Routinely, when I would have a seizure, it would be followed by a migraine headache. This is because of the arteries in the brain condensing and expanding. I did not have a seizure from the stress, but it had to do with the internal bleeding that may have been going on as the muscles in my leg just locked up.

I did not have a headache – but my left leg was about as heavy as a brick cinderblock. As I woke up, the first thing that I noticed is that my entire left leg is numb. Apparently, my seizure had been caused by stress brought on by my heavy breathing, and that was enough to smash the left femur into three separate pieces, which in turn, caused the other muscles in my leg to clasp together. When my hip fractures, the parts of it disrupt the blood supply in my leg. I woke up from the seizure on my kitchen floor, with no headache, but instead, had a left leg with absolutely no feeling, that was scaring the living hell out of me. Not only was my leg asleep, but the blood supply was not returning.

For the first time in my life, I thought that I was going to have to have my leg amputated. I was not thinking about the pain, but more or less, about the very real possibility that I would be spending the rest of my life confined to a wheelchair. My life up to this point was already complicated enough. The one thing that I didn't need was to have to submit myself to the care of family members because I was unable to support myself financially in any way. My parents were helping me with my rent while I was able to cover my other bills such as my car payment and insurance – with varied amount of difficulty at times. The next two days would be spent playing a high stakes game of poker deciding whether or not I was going to seek medical treatment. Perhaps this will just go away on its own and I won't have to worry about making major changes at least for another few months. Well, I guess bluffing won't make a problem of this magnitude disappear off the radar screen. Maybe I should just fold.

I would rather have the damn headache. Because, waking up from a seizure and finding your leg asleep with no sign of getting better is scary. At this point, I pray to God and ask him to spare me my legs. When I tried to get up and walk on my leg to make it better, to my surprise, I find out that it is impossible for me to stand on two feet. I did not want to spend the rest of my natural life in a wheelchair, rendered useless to society. I tried to attach some sort of meaning to this suffering. Eventually, I find out what my real job is going to be doing. I know that I will eventually have permanent job stability – but I don't even stop to think about how much my story is worth. Attempting to resist the process of change, I walk on my leg and find out some other information with regard to this injury.

Knocked out of my routine, I decide that I am going to just walk around on this leg a little bit. Exercise is good for a pulled muscle, anyways. Zap! As I try and walk on my leg, I feel like I am being electrocuted as impulses shoot up my leg with even the slightest contact with the floor. Zap! I still think that there is something unknown that I have to find out about. Holding the railing of the staircase for dear life, I try to walk down the stairs and realize I am in serious danger. If I try and hop down the stairs, I could misjudge the distance between the stairs and land on my back. If I walk normally, I am electrocuted. I am damned if I do, and damned if I don't. I decide to do what I really should not do, which is walk on the affected limb.

I was in for a shock. Upon trying to walk on two feet, I felt what was similar to electricity firing up my leg. It is so painful, that when I am trying to fix the situation, I grab onto the side of the apartment complex when I am walking on the grass, grasping the side of the building as tight as my cat grasping my supply shelf in my office area of my apartment. I felt like Spiderman clinging to a wall. It must have looked comical to a few people who had absolutely no idea what was going on with me. I didn't even know what was going on with myself – and I was afraid to find out. Afraid to tell people what I am experiencing, I start popping pills like there is no tomorrow. I take about two bottles worth of pain medication – which is enough to delay my surgery by about a day while they spend time flushing the medication out my entire system. As a result of doing this – and delaying going to the hospital – I end up running an extremely high temperature. As I am having this done, I am chastised for making a bad decision by someone who works within the "system". I probably had a seizure from the stress that my body was under. When I was in the hospital, my temperature was 103 degrees, which eventually cooled down to normal body temperature with a day or two. A few more days of fighting the devil, and I might not be here to tell my story to you.

I am hoping that this pain goes away so that I can get to my job for my temporary agency on Monday, which is a job that I have during the day. My pain goes away in my leg. But the electric shock treatment continues if I walk normally on my foot. Why in the name of God do I have electrical surges shooting up my leg? I have never experienced anything like this before in my life. Unless I stand on my tiptoes, there is no way to avoid getting zapped. Even touching the floor *slightly* with my foot, putting no pressure on my foot at all whatsoever, causes me to clench my teeth in pain. There has to be some explanation for this. Why would I have electrical charges shooting up my leg, even though the swelling is gone?

That is another job that ends. It is as if this is a revelation that I am supposed to be doing something different with my life. Somebody is sending me a message to do something differently. As I go through two bottles of pain medication, it is clear that nothing is happening to clear up the pain that I am experiencing. Out of all of these situations, pain medication never ails what is causing my problems. What used to be a symptom has now turned into a major problem.

Even the slightest amount of pressure of any kind against the floor will result in thunderstorms happening in my feet. When I was not able to make it across the parking lot, I struggled, leaning up against my car while I was standing on my one good leg. I moved my car closer to my apartment. I know the time is going to come when I have to make a dreaded decision before I lose this hand of poker and get dealt the death card – the one that will signify the end of my life. It wouldn't take a psychic to tell me this. A trip to the hospital was imminent.

I try everything. Stretching exercises don't do the trick. I resort to applying cold water to my leg. The cat, during the time that all this is happening with me hopping on my leg, follows me around, thinking that I am playing some sort of game with her. That cat has a death wish. In the process, cold water gets dumped on the cat – oh well. That cat has four good legs. At least she can run. It gets to a certain point when a critical decision has to be made. I decide to headquarter in my reclining chair and order pizza. As I balance on one leg to get my pizza, sign the credit card receipt, my cat decides to try and run out the door. I am in no position to kick the cat back in the apartment. The less I have to walk on my leg, the better. At this point, my mother is going crazy. I have not called for about two days now.

Deciding to seek medical treatment is a decision that I am not going to like having to make. But if I want to ever walk again in my lifetime and live something resembling a normal life, I am going to have to take it into consideration. I don't know if I am going to get dealt another round of cards that will unequivocally change not only my life, but result in close friends and family members talking about me in the past tense instead of the present tense. I was unknowingly running a very high fever with no way of knowing why. Everything was fine before I went to deliver the telephone directories.

The next morning, the pain is not noticeably better than the previous night, even though I had been applying a cold washcloth to my hip area. At 7:45 in the morning, I hobble down the stairway, holding onto the railing for dear life. Walking on the grass, I slowly make my way out to where my car is parked. I get behind the wheel of my car and drive to the hospital to turn myself into the authorities. As I drive to the hospital, I have difficulty applying the clutch. As I drive to the hospital, a lake starts to form in my eyes. The amount of pain is so intense, that I have tears streaming down my face.

As I will find out, my problem is in the hands of the medical industry. But the journey only gets more difficult, especially when I have to be the one to explain why I got a fracture and how I got it. Hopping on my right foot across the parking lot of the hospital, I hope that the problem that I face will be one that is temporary and can be fixed. I am thinking that at most, I will only be in the hospital for about a day at the absolute most while everything is being taken care of. I am thinking that perhaps I will be given stronger medication so that I can get back to my everyday life. That never happened. I was yanked out of that everyday life, and was seen as somebody that was completely different from everyone else. I had no way of knowing it, but life would eventually change for the better for me. It certainly hasn't changed yet. And I have not even seen my movie optioned further into development by any of the producers that have my script.

THE IMPENDING STORM

The emergency ward was very quiet. The morning was about to become a lot more intense upon my arrival. When I check myself in at the emergency room, I am asked what the problem is. That is when I lie. I tell them what I think happened to me – that I pulled a muscle. With no other information, that is the only information that I could give to them. I didn't know what really had happened to me. The critical thing, was that I did not come to the hospital the same day that I was loading the phone books. The problem, was that I did not even know that I was dealing with an injury at that point. It was probably assumed since I was not referred to the hospital by the person I was working for, that this injury was not work related, even though it all started to develop when I got home. That is where it gets complicated.

As I have my hip x-rayed, I am informed that I am going to have to be operated on, but it won't happen until the following morning. I am informed that the ball and socket joint of my hip has fractured in three different places. To this day, I know that to be the case, especially if I lean on my left hip in the wrong way because I can feel where the titanium pins are inserted so that I can walk like any other normal person. If I cross my legs in the wrong way, I notice pain in my hip, not to mention the fact that I am cutting off the blood supply to my legs. I tell myself another lie. I believe that I will be on my feet like normal, as though nothing ever happened to me. That is where my struggles really begin.

Come on, now, wake up – is anything in life ever that easy to deal with?

Chapter 4

Surviving the Hospital

The hospital was quiet before my arrival on April 2, 2002. Just a few days earlier, I had been through a test of the wills that nobody in their right mind would ever expect to go through. Out of pain medication, and most of all, out of patience, I walk – with a very painful grimace down the stairway of my apartment complex, doing my best not to make it obvious that there is something wrong. Gingerly walking on the grass up to where I had moved my car some time earlier, I lean up against the side of my car, fumble in my pocket for my keys, and open the door. As I go to the hospital, I am put through such pain that my eyes well up with tears. This was only the beginning of my journey. And it would not be over until I got out of the hospital so that I could tell my story to the entire world. Although, it would not happen that quickly. I would be facing a lot of road bumps along the way to financial success – but I was determined. I was determined to change a system that was screwing up everybody's life in one way or another.

As I arrived at the hospital, and people learned of why I was admitted for surgery, there were more people interested in what the hell I was doing, at twenty-six years of age, in a hospital room with a fractured hip. It would be this trip to the hospital that would answer the question that I had been asking for over twenty years – I would finally find out why I got epilepsy. What I would not know, is the people who I bonded with in the hospital – my doctors and my internist – would be upset with my analysis of the situation.

Oh, I remember what I was doing at the hospital. I would be teaching the medical profession that everything they were taught in medical school would eventually be questioned. As they don't know how to appropriately diagnosis my situation, it becomes clear to me that I am different – and I personally like it. Ending up in the drunk tank for a second time, I realize that there is something else that I should be doing.

As I am given a lot of thinking time, I start to piece together the events. At the same time that my decision to try and lift those phone books into my trunk required sacrifice on the part of my parents, it would be paid back with high dividends as I became a teacher, educating people about the dangers of long term anticonvulsant use. Getting to that point would not be easy.

On the date of April 2, 2002, I made the decision to turn myself into the authorities. Just like a scofflaw, I had been running from something that I could no longer hide from. Fighting was futile. Running was scary. Hopping through the emergency room doors, and check in at the front desk, standing on one leg, clutching the desk in the admitting area for dear life. Filling out the form, where it requests to know why I am here, I say that I pulled a muscle. Essentially, I told a white lie, because I couldn't imagine what else I could be suffering from. It seems like years before someone comes into the waiting room with a wheelchair. Now, I was going into the hospital with the assistance of a wheelchair – and this time I would come out of it a completely

WELCOME TO YOUR NEW HOME

Since I had nothing to do, I call my apartment to check the messages on my voice mail. Call it bad karma, but as it turns out, the day that I am admitted to the hospital is the day that I get a call from one of the temporary agencies that I had been following up with over the past few weeks. The message said that the client that he had sent my resume to liked it, and said that he had a temporary assignment for me. I don't know what it is with hospitals and employers. Every time I am near a hospital is the time that I find out that there was a job out there for me.

Actually, this is an interesting way of looking at this. Because it was around this time that I realized that life as I knew it was going to change permanently. The prosecutor walked into my room, asking me how I my injury had occurred. From the tone of his voice, it was as though, by the way he asked his questions, that he was doubting my story. I knew what my rights were. At this point, I didn't even have an attorney representing me. Remaining true to my convictions, I stuck to my story about how, for some unexplained reason was unable to walk due to a fracture, which was not present before I took the job to deliver telephone directories. When I told him my story, he didn't believe me. I had a long battle ahead of me. Doctors were just hearing of my scenario around the same time that my surgery was being performed. In essence, because my case was very unique, doctors had difficulty coming up with an accurate diagnosis for my condition. My internist did not know whether I had full blown osteoporosis, or osteopenia, which is a milder form of osteoporosis that I could recover from with proper medication.

The reason that my case was unique to the medical world, was that there were no research studies about the role that anticonvulsant medication played in relation to a person's bone density.

The osteopath told me that based on my age, it would be better if I had titanium pins implanted instead of a hip implant which wears out every few years. With my age, the titanium pins were better since hip implants are used in people who only have maybe ten or fifteen years left. At this point, I really did not care what was done. All I wanted to do was get this surgery done, and get out of the hospital. What I would find out was that it would not even work out that way.

. As I was preparing for how I was going to handle the surgery, I had an order of business to take care of. And this was not going to be easy. I had to find a way to break this news to my mom. Never mind. I bet she already knows. After all, I haven't even been returning her calls for the past two to three days. Plus, I figure I will be out of this place after my surgery. Hold on one minute. Something is telling me that I am going to be gone for longer than that. Even if I could go home, I am not going to be able to walk on my leg until the blood circulation to my hip is restored. And I can't drive my car since I will have to put the clutch in to shift gears. How in the hell is life ever going to be the same anytime soon? It never will be the same, it will only get better – when I find a way get paid big time for everything that I went through. And there will be a whole lot of people who think that I am a nutcase. Later on down the line, I would eventually end up staying within the four walls of my existence while I figured out a way to escape from my stomping ground and go to great lengths to explain why I decided to move two thousand miles away from home.

I did not have my medic alert neck chain on that day, and that is when I realized that I was in serious peril. Without the medication that I needed to survive, things were surely going to get a lot more complicated. When the nurse said that my mother is overprotective, I imagine that she was referring to the fact that my mother had to give them the details of everything that I was taking for my epilepsy – and probably asking too many questions with regard to my prognosis, which nobody would be able to answer.

I did have company visit me occasionally in my room. Over the next two days, I had a lot of social workers visit my room. My big concern was finding a way to cover my expenses. Something that never occurred to me, was to file a worker's compensation claim. Since this was an on-the-job injury, I could have my medical bills covered by worker's compensation. Proving my case was going to be difficult, as I was only there for the one afternoon, and did not even have the names of anybody there. All I could give was a vague description for who was present during the time I was loading my car up with phone directories. However, explaining to the average person, that it was the lifting of the phone books that caused my hip to fracture would prove difficult. How do you go about describing how heavy the load was? I know how heavy my Forest Cat is, because I picked the creature up all the time with no problems whatsoever, so I used that as a guide and said that I thought that the phone book bundles, which I could not lift with my hands, weighed fifty pounds or more. My cat weights something like twenty pounds, which gave me good reason to believe that the phone books weighed fifty pounds.   In all the years that I have moved around from one location to another, I have lifted things with varying amounts of weight – but that paled in comparison to these phone book bundles.

A VISIT FROM A FAMILIAR FACE BACK HOME

As I was sitting in my hospital room, reading a magazine and flipping channels on the television, I get a visit from someone who looks all too familiar. As a matter of fact, it was almost like I was in déjà vu, stuck in a time warp. What was he doing here? He looks so familiar that I lock my gaze on him for about thirty seconds to a minute straight, trying to jog my memory. The longer that I look at him, with my eyes about as big as dinner plates, I resist the urge to drop any hints that I might know him – like the huge smirk that is forming across my face as I stare at him. This person looks way too familiar. Am I seeing things? Am I dreaming? How could this be? Then he introduces himself – saying that his name is Chris. I know that I recognize him from somewhere. He was the head supervisor of the billing department back at the hospital in Milwaukee that I used to work second shift at. I wondered what he was doing down in this neck of the woods. Of all the chaos that was currently going on in my life, he was the one visitor that put my fears at ease – at least temporarily. At least he answered my questions with regard to worker's compensation. I was getting the brush off whenever I brought up the issue, since nobody here seemed to give a shit about my welfare. Since I had supposedly known him from the northern country under different circumstances, he was more than happy to help out answering my questions. He seemed like he was a little too helpful to be anybody who was directly affiliated with the system that I would be dealing with.

To see if I was right on my assumptions, I listened for if he would stop at the nurse's station as he left my room to possibly see who I was. When I heard him, supposedly ask the nurse who I was and why I was in the hospital, she said that I was lifting telephone books and ended up fracturing my hip, something along those lines. When I heard him stop and ask for more information, I knew that possibly I was dealing with the same person. I start to try and grasp at straws, to think of whether he had been known to make any business trips possibly to the southern part of the United States. I remember when I was working at Aurora Healthcare, he would be gone for a few weeks at a time, down in the Texas area for different purposes or in different regions of the country for business related purposes.

Perhaps I am not in dreamland. This could very well be the same person. I am sure picking up a weird vibe that I normally would not be picking up from a complete stranger. This is just way too weird to be seeing this same face in a completely different setting than I am used to seeing him in. I didn't want to completely blow my cover, so I bought time by asking him additional questions as to what I needed to do to get my worker's compensation claim started. All the questions that I asked, he seemed really helpful in terms of answering them. With all of the faces that were coming and going from my room, his face was one that brought a vibe of serenity and peace to me – something that I very much needed in this time of chaos.

I started to fixate on this more and more, so I came to the conclusion that this was the same person who I did work with at Aurora Healthcare in Milwaukee, Wisconsin. I hope that if he sees his name in print, that he will tell me whether or not he was in Tempe, Arizona on the date of April 2, 2002. I am going to venture a guess that the person who I saw in Tempe St. Luke's Hospital on that doom-filled day was Chris Peters. If I am right on my assumptions, feel free to send me some sort of sign through my publicity agent or through email. If you are who I think you are, then we definitely live in a small world. Moving to a city with three million people and the fifth largest city in the United States with close to three million people, I never would have expected to cross paths with him after moving from Milwaukee. I guess you learn something new every day. I guess there are people in this world that are concerned about my welfare.

THE SYSTEM

Wheeled into the hospital room, I am bored out of my mind. If I knew that I was going to be spending a long period of time here, I would have brought reading material with me. Aside from the time that I spent out of my room having various tests done one me, I used the television as a way to pass the time. Then, an osteopath, who would later perform my surgery, walked into my room to ask how my injury occurred. I told him that I had been lifting phone books when this happened. What I did not know, was that this doctor took a lot of the worker's compensation cases that resulted in injury. Something else that I did not know, was that the worker's compensation carriers would use him fairly often because he would say exactly what the insurance companies wanted him to say, in effect doing complete injustice to the court system. If it were not for the fact that I retained an attorney later on that dealt with worker's compensation matters, I never would have found out that piece of information.

Titanium pins would serve as a reminder of how I am different from most people. What I did not know was that as a result of having these titanium pins implanted, there would be times that if I was sitting the wrong way, that I would feel pain from the pins, which ran through my upper femur into the ball and socket joint of the hip, which I shattered during the time that I lifted the phone books if I leaned or sat in the wrong position. As a result, I would usually try and prop my left leg either up on the wall or I would extend it out.

I agreed with my osteopath's suggestion. Whatever he could do so that I wouldn't have electrical sparks firing up my leg again. "Just put the pins in so I can get the hell out of here", was all I happened to be thinking at the time. This stay in the hospital was worse than the injury itself. My surgery was actually delayed because of my high temperature and the fact that they had to flush the medication out of my system. What would be involved would be a lot of noise associated with the pounding of titanium pins through the fractured part of my hip so that I could walk without feeling pain.

The following afternoon, I was taken down to the operating room, where my osteopath gave me anesthesia, and I drifted into a peaceful sleep for the next two hours. Slicing the upper part of my leg open, he felt around for the three bone fragments that were once a part of my hip. There were other people who were making sure that I got the right medications for my seizures so that I wouldn't have a seizure during the time that I was in surgery. The surgery would last about two hours as the titanium pins were pounded into place.

Waking up in the recovery room, I was still groggy and fighting the anesthesia that had been given to me. I needed to wake up quickly because it would not be long before I would be fighting a much bigger system that really did not care about me – much less about my creative endeavors when I decided after losing my health insurance to shift my energy to working in the entertainment industry so that I could at least pay for my medications without adding more to the national debt in this country. Because Arizona is one of the few states that does not have an adequate high risk program for people that are not able to get insurance through normal means, that served as ammunition that I needed to challenge why it was that I had epilepsy in the first place. I figured, now was as good a time as any other to be asking why it was that I had epilepsy. This would be the next family rift that I would create as I declared that vaccinations were absolutely responsible for my epilepsy. After all, someone out there somewhere had the nerve to do the research to make me think this.

When I was taken up to my room, I slept most of the day due to the anesthesia that was in my system. I never realized how much I would be drained just by having surgery. Before I was transported to rehabilitation, the hospital gave me a pair of crutches so that I wouldn't have to use my left leg to get around. Until there were x-rays that would show that my body had not rejected the pins, and that blood flow was restored to my hip, I would have to get around using crutches. At least going to the hospital was positive in one aspect.

REHABILITATION

My parents would not be able to get out to help me for at least another week. Trying to arrange for time off work, they could not just leave on a moment's notice. I was currently stuck in a place where I was the youngest resident – all the people who were in the rehab facility were using canes and walkers to get around. I didn't feel like I belonged here. I felt like I was living the life of an older person who was confined to a nursing home due to old age. Was I simply a young person trapped in an older person's body? Was that what my life was really coming down to? This must be some sort of bad dream.

A lot of the residents in the rehab facility had artificial hip implants as a result of a fracture. As much as I had something in common with them, I was different, again – age wise. I was also fair game with people constantly asking me why it was that I was on crutches. I was starting to get tired of the circus and starting to consider finding a way to get the hell out of the hospital.

Given the time to do some additional thinking, this was when I was making a mental note of the connection that epilepsy has to osteoporosis. It was the only plausible explanation, given my age and gender. Even before the bone density test was ordered by Dr. Frank Metzger, my internist and primary care physician, common sense told me that at my age, the only reason I would have had a fracture was from osteoporosis.

A lot of people who are diagnosed with osteoporosis are female and past the age of menopause, when the body is no longer producing a normal amount of estrogen. In my case, hormones were not much of an issue, although I did get a referral to an endocrinologist to make sure. Since men produce testosterone their entire lives, what would make a twenty-six year old male show up at the hospital with a fractured hip? That is when the reality really hit me hard. With epilepsy and osteoporosis, now I was really different and more people were finding that out, as I decided it was time for me to speak out against a system that was responsible for causing so many problems in otherwise healthy individuals. Due to the fact that I had no way of getting insurance through normal means, I had to find a way to get rid of the epilepsy. The opinions that I held would make me unpopular with a few people who perhaps didn't have a lot of information about holistic medicine.

With the chance that my hormones were out of balance were minimal, I reasoned that it could only be one thing that has caused my hip to fracture. The medication that was keeping me alive and keeping me from going into status epilepticus, which is deadly if medical attention is not found immediately, had caused me to have a fracture. I had moved around a lot in my life and I figured that I would have noticed a problem a lot earlier. Why is it that I notice this problem now?

Fear was overtaking me as I was talking to my mother on the phone. At this time, I did not know what my diagnosis was going to be. That was the part that was truly scaring the hell out of me. At a certain point, I was scheming a way to sneak out of the hospital, even though I had three titanium pins through my hip. The only problem, my car was fifteen miles away at another hospital and to make matters worse, again, I am in a situation of not knowing where I am located. Since I figure that I have nothing to lose, I verbalize, real loudly, so that anyone within earshot can hear me, that I am considering just walking out of the hospital if my parents don't get here on time. It certainly ruffled a few feathers. I had to find a way to pass the time somehow – and flicking channels on the remote control was only going so far.

Since I couldn't get my cat all ruffled about by creating drama, I decided to try the same experiment on those around me. I was bored and just needed to create some ruckus. Every minute seemed like an hour. Every day seemed like it was an entire year. Every day seemed like it was a life sentence. I told my mom that I might as well be serving time in jail. I knew how I could quickly sneak out the back door – if only I had a way to get back to where my car was – but it was not worth it. As they say, patience has a virtue. By waiting it out, I would not have to worry about my worker's compensation claim being denied. I wanted to stick somebody with a huge expense – and had it not been for that, I might have taken it into consideration. But sticking somebody with that huge expense would be very difficult to do – but would definitely be sweet revenge that I looked forward to. Besides, where the hell am I located?

THE BONE DENSITY TEST

As I was talking to my mom on the phone, a doctor came into my room pushing a wheelchair. I was already being pulled in numerous directions and wanted to know what this involved. Dr. Frank Metzger, my internist who I had not yet met, had requested that I undergo a bone density and bone scan with the uniqueness of my situation – the very same tests that should have been done before I had surgery at the hospital. These tests would confirm what I had been hypothesizing for the past few hours before I had these tests. I had no idea how my life was going to change. For the first time in awhile I did not know how I was going to make it out alive. Was my life going to change or would I still be doing work in relation to the field that I graduated in just six months ago? These questions would all be answered when I had more time to reflect on what happened to me and why it happened.

As I stared up at the monitor, which was displaying every bone in my body on a monitor, the worst things were coming to my mind. What if my spinal cord was really fractured? Nobody was telling me what was going on. I had to spend the next day wondering what was really happening. Why had somebody that I had not yet met in person order me to have bone scan and a bone density test? Why was it that this was not done before I had the surgery? These two questions are the pivotal questions that outline what was really never told to me.

LIFESTYLE CHANGE

As I am put in the wheelchair and taken to have a bone scan and bone density test performed, I go along with the flow of things, with full knowledge that life as I know it is going to change forever. My plan is to get the entire world to listen to my story. It was my personal foundation and long time friend, who told me that advances paid to writers are not that great. What I am completely unprepared for, is the euphoria that will result when I find out that there exists a possibility of a seven figure income as a writer for Hollywood. This is something that consumes my life. Because by having a movie devoted to my story, which will no doubt increase the value of my advance that I eventually will receive for my book. I somehow had the inclination to believe when I left the rehabilitation facility, I was going to be known by millions of people. What I was unprepared for was how these millions of people were going to react to something that was completely unheard of from a medical standpoint.

What I did not know, was that my vision would not be realized in a way in which I expected it to be realized. The entire reason the movie came about was because of the work of a con artist, a dishonest literary agent who took money from me upfront telling me that my work could be taken by a publisher. I believed her. I knew that my story was unique. I knew that there had to be a publisher out there somewhere that would kill for this story to be in print with their name all over it. What I was completely unprepared for, was the fact that a lot of publishers made the decision based on the title of my book that I did not possess the credentials to discuss that which the doctors were unable to understand.

It was the fact that I did not have a medical degree which told publishers that I had no business telling my story to the world – don't worry about the fact that I almost killed myself to get to this point in my life to try and make a difference in the world. I think that due to the fact that a lot of them had never heard of something like this, they were questioning me. Hopefully, after my movie is out and my name is seen on the ending credits, and I have been giving interviews on television, that might convince some publishers that I do have the credentials that are necessary to tell my story.

THE TEST RESULTS

The following day, Dr. Metzger walked into my room, with results of this test. He asked me what medications I was taking for my epilepsy, which was when it really hit home. A medication that I took as a child had caused this to happen. Evidently, the fact that I was taking Tegretol for fourteen years, was the primary reason for why I was in this situation. The end result was secondary osteoporosis with a bone density which was fifty percent below normal. A normal bone density for someone my age would have been 0 or 1, according to hospital records, my bone density registered at –4.6.

According to a report that was published in *Journal Watch Neurology,* it took a sample of 59 adults who were on some of the earlier anticonvulsant medications, which include Tegretol, Dilantin, and Phenobarbitol. The key things about these medications is that they are enzyme producing medications, which block the body from being able to metabolize vitamin D. Out of these adults, more than half had a condition called osteopenia of the hip or spine, which is a forty percent loss of bone. As it came to my attention, there was something else far greater than this that commanded my attention. It was going to make me the black sheep in the family as well.

In a statistical sense, I would be considered an outlier on the data report. This information would not even be within the normal range of what is considered normal bone density by any means. I felt that due to the fact that I was outside of the ninety-five percent of the sample population, I would be a voice that would be heard when it came time to educate people about my life story. What I was unprepared for was the fact that, due to the fact that I lack a medical degree, getting somebody to listen to me would be anything but easy. It only seemed like insult was being added to injury when I was told I didn't have the background necessary to tell my story, never mind that it truly happened to me.

Dr. Metzger asked me if anything weird had happened to me in the past few months. This was truly the defining moment, as I now knew that what had happened to me only six months earlier was not related at all to my car accident. I told him about the time that I urinated blood three months ago, but said that I thought that was in regard to my car accident. Hearing about that incident, he asked me further questions about my medical history, specifically about the medication that I was taking to control my seizures. I had been on numerous medications, and it was impossible for me to remember them all. The one I was on the longest, however, was Tegretol and I took that for fourteen years until I switched to taking Neurontin and Felbatol. I was on many other experimental medications, ones that I could not remember. My mother had a better memory for that, since she was responsible for my care during that time. To be honest with you, if I had not been vaccinated as a child, I would not have been on *any* of those medications. Dr. Metzger told me that he was going to have somebody in the pharmacy department do some research on some of the medications, specifically the ones that I have been taking. In his professional opinion, he thought the only explanation for my diagnosis of osteoporosis, given my age, was the fact that I was on a lot of prescription medication as a child. This was the same thing that I was hypothesizing even before the test results were run.

FOLLOWING THE LIGHT

I moved to Arizona to get away from the demon, only to find out that it was going to find another way to nail me. Dr. Metzger wrote a prescription for Actonel and gave me specific instructions on how to take it. I had to take this medication on an empty stomach with a full glass of water and could not lie down for twenty minutes afterwards. He also advised me to take 1500 milligrams of calcium fortified with vitamin D every day. The vitamin D would work with the Actonel to build bone density in my skeletal system that had been lost, as well as prevent any more fractures from happening.

Otherwise, if this medication came back up my esophagus from abdominal fluids, it could mean that I would have difficulty swallowing. I already have enough problems, and did not need any more problems. That was the least of my problems, however.

The scariest part of the test results was not the fact that my bone density was low, but that my bone scan was showing that the lumbar part of my back, located near my rib cage, was showing evidence of an internal fracture. My left heel, where I had occasionally had pain during my growing years was showing serious activity. I was going to find out just what this was going to mean when I got home. Confined to the use of crutches, I would not know the full extent of the risks associated with my low bone density until I started to walk five months later.

Instructed to slowly wean myself off the crutches was a chore. I could put weight on my leg, but I had to do that with the support of my crutches. I was also instructed to do some type of weight bearing exercise to get regular blood flow back to my hip. My mom always complained about the fact that I carried this heavy computer bag with me most places that I went. Just walking up the stairs with the computer bag on my side, was excruciating. Concerned that doing this would re-injure my hip, she told me not to do this. However, it is my belief that by doing this, I actually helped my situation. When I went to the final appointment to see my osteopath, also the attending physician who performed my surgery, for the first time I was actually able to see blood vessels which lead to my hip on the x-rays that were taken. A number of times, the x-ray did not show anything with regard to blood coming back to the hip. If blood flow did not get returned to the hip in time, then the head of the hip would fall off, which would have resulted in the need for a hip implant.

DEEP VEIN THROMBOSIS

A little bit of time went by before I was instructed to put a vibrating cuff around my leg and keep it on during the time since I was not using that leg regularly. There have been reports of people who are off their feet for long periods of time that have developed clots in their leg, which is known as DVT (Deep Vein Thrombosis) and died as a result of that clot traveling up to their lungs or heart. That was how the price of poker was raised for me during the second time that I was left without an ability to walk. Little did I know that just by not going to the hospital for two days, and trying to deal with this injury myself could have been a fatal mistake. In addition, when I got to the hospital, I was already running a high temperature. By going and getting medical treatment, I essentially spared my life and avoided getting dealt the death card.

Since I did not have my medications with me, I did not consider that there were things that were going to be affected because I was in the hospital. Never mind the fact that right when I was admitted to the hospital, that I happened to check my voice mail and there was a call from one of the temporary agencies I was registered with, who was calling to let me know that they had a job for me. I called and told them that I was in the hospital. I don't know what it is with hospitals and jobs for me – every time I am about to find a job, I am somehow confined to the hospital. Maybe this is just bad karma. One of these days, I will get a job.

I will get a job when I decide to employ myself 24/7 around the clock writing a movie script. With my phone not ringing much these days, I had no other choice. After my injury, I decide, against my mother's wishes, to sacrifice everything in order to educate the medical establishments about what they don't seem to understand. Putting everything out on the table, I thought that I would be doing the entire world a favor, not to mention the three million people in this country who have epilepsy, to tell them about my story and how it relates to them. Even those who don't have epilepsy would be able to find some sort of meaning in my story. Unfortunately, there were those who did not see eye to eye with me. Despite the fact, according to my attorney, that my script is high concept, convincing a studio to take a risk with hundreds of millions of dollars simply is not an easy task.

Presently, I am literally depriving myself of sleep, working around the clock to get my business off the ground. If the time stamp on my emails says anything, it would spell sleep deprivation. There were times that my body would just go into "sleep" mode, and my cat would come and wake me up – with her loud mewing. Her mewing would be loud enough to wake up the dead – especially when she did not have any food to eat. If I did fall asleep, she would be sure to mew at a loud noise level until I woke up – only to find out that nothing was wrong. She just didn't like the door shut when I was sleeping. I did that because I did not want my feet nibbled on while I was sleeping. Maybe I should just feed the monster so that it leaves me alone.

GETTING NOWHERE

During the time that I was trying to get my worker's compensation resolved, I was getting nowhere. What I did not know, was that I was placing my trust in people who were obviously a players in the system. They couldn't have cared less why I was injured. It was their goal to tell me that in their own "opinion" that my injury was not even related to the work that I was doing when I was trying to lift those heavy phone book bundles. When I filed my worker's compensation case, it was denied. I ask my osteopath what his "professional" opinion is with regard to my injury – and he claimed that my injury was from "normal everyday activities" in those words. I couldn't believe my ears. What the hell was going on here? I was at the mercy of somebody who was judge, jury, and executioner all in one. How could he draw this determination when he didn't even run a bone density test? Whatever he said the injury was from, was what the court would believe. There was nobody who could cross examine him – until now.

Remember, I emphasized that I had been lifting telephone books, which directly played a role in my fracture. Every time I had follow up appointments with him, which I had to go to, or I would jeopardize my case, I could feel a vibe in that office of extreme anger on the part of his assistant. She was almost like a drill sergeant. I had something faxed to his office with regard to the fact that due to the fact that I was on crutches, I could not work and wanted her to sign it, so I could at least collect unemployment compensation. She was not only going to be a complete liar on the part of the insurance company, but was also going to be extremely unhelpful in regard to anything that I needed done. If anything, all that she did was the bare minimum – and then there was that vibe of anger that I was picking up from her, like "Why am I wasting her time?" Her attitude was, "Why be helpful when that could end up hurting me somewhere down the line?"

She claimed that the form never got faxed over. I stood there and watched the form go through the damn fax machine. She is claiming that she never received it? I was starting to hate these monthly visits to my osteopath's office for routine x-rays of my hip see if there were any signs of improvement. I would rather be going to the dentist to get my fillings replaced. Every time that I would get x-rays done, I would not see anything on the x-ray. What was the point of these monthly visits?

Oh, I forgot, there was a chance that blood flow wouldn't come back to my hip, which would result in necrosis, where the ball of the hip crumbles and I would have to have the surgery redone with an artificial limb. That was the least of my problems right now – all I want to do is find a way to survive. Try doing something like that on cancelled health insurance. That is not cheap by any means. My operation alone cost about $50,000, including my room and board at the hospital.

I had nothing to lose at this point. I started a verbal fight right there in the office demanding to be heard, raising my voice. I usually hold myself back, but in this situation, I was so pissed off that I didn't care what would happen. I knew that she would not be dumb enough to call the police, because I would probably have revealed some things to them that she did not them to know.   She would have none of that, and basically told me "No,no,no, you don't do that!", talking down to me like a little kid – and then proceeded to turn her back to me like the coward that she is. By her mannerisms, I figured there had to be something behind this. I couldn't figure out why she was doing this. Now, I knew that she also was a player. I couldn't put my finger on it, but somehow, I knew that she was involved with the system.

Even without any definite proof, I knew that she was an assistant prosecutor – and only existed to make my life more difficult. She even tried to put words in my mouth and told me "When you were in the hospital, you said you were unemployed!" I have been used to sticking up for myself in different situations before – playing the role of being my own attorney – and knew the damn game that was being played. If she could get me to say that I never worked for the company, then legally, the company would be off the hook for having to pay any type of worker's compensation benefits. I had my father call the company and make sure my name was still in the computer as an employee – it was. Nonetheless, she had no clue who the hell she was dealing with. This all boiled down to one thing, which was protecting her job – as well as getting some extra kickbacks from the district attorney.

If there was one thing that my father would do that would help out my situation, it would be to put this bitch in her place. If she thought it would be fun to play with my emotions, my father is nobody to mess with on a bad afternoon. My father was somebody who would stand behind me at any and all costs, especially where my life was on the line. Dealing with me was like a stroll in the park. I just let her kick her heals in. Her attitude would come back to haunt her when she least expected it. Dealing with her was worse than having the x-rays taken of my hip.

When you are all doped up on anesthesia in the drunk tank, you tend to say a few things without necessarily knowing in what context the other person will intend for those words to mean. How convenient of you to twist my words into your favor. Yes, lady, at that time I was unemployed – because I couldn't work due to my hip fracture. What a God damned bitch.

COURTROOM DRAMA

I was fighting a very big system that couldn't care less about me – and I was at my wits end. All I wanted was to have life return to normal like it was before I moved from Milwaukee. Something told me that this was never going to happen. I knew that I was getting nowhere with the assistant prosecutor and the district attorney every time I went to have my hip evaluated. After I got back home, I grabbed the yellow pages, and looked for an attorney that dealt with worker's compensation. I needed someone to go up against these jerks and put them in their place. Something did not seem right, and I knew that I couldn't do this alone. What I would find out petrified me. My attorney called to check on my claim, which had been denied, as most worker's compensation claims are, and he found out that there was never an accident report submitted by the hospital. Generally, a report is filed with the industrial Commission of Arizona when there is a situation where a work related injury is involved. But, if I said that the injury was work related, then my private insurance would be off the hook from paying out anything, because it was expected that the cost would be picked up by the employer's worker's compensation. When worker's compensation denied my claim, everything had to get resubmitted to my private health insurance company.

When my attorney called the worker's compensation number, he was unable to find any information with regard to an accident report. Did I not tell them numerous times while in the hospital that my accident occurred when I was lifting telephone directories into my car? Are the people at the hospital deaf and stupid? Somebody is playing a sick game of cat and mouse.

I guess they forgot about that part – very conveniently. The worst part, is that the one person who I felt could put in some words on my behalf did not feel that they were a medical expert. A lot of that has to do with the fact that he came on the scene after I was transferred to the rehabilitation facility and had nothing to do with the surgery itself. Even though the bone density test was ordered by my internist, he said that all decisions were deferred to the attending physician who had performed the operation. My osteopath, without conducting any tests, lied by saying that my fracture was from *normal everyday activities.*

My osteopath had taken the liberty of deferring to someone else the duty of scooping out the cat's litter box. I suppose that is the entire game here. I felt like that was a conflict of interest. How could you be one to have an opinion about how something happened, when the opinion you give will be one that makes sure that you have dinner put on the table tomorrow night? It made absolutely no sense to me whatsoever. Aren't courts supposed to be institutions that enforce peace and order and truth on all the parties involved? That is the one thing that absolutely puzzles me. My attorney even said that if it were to go to trial, the insurance company would call my osteopath, knowing that he was going to tell a lie about what really happened to me. Why would he tell a lie? Because if he says something that the insurance company does not want him to say, he won't get as many cases referred to him. It is a bad system and I don't see too many people doing anything about this type of stuff. It all has to do with the hand who feeds him.

The hospital knew that this injury was work related. They also decided not to run a bone density test to find out why I had fractured my hip. What do they do? They assign me to a doctor who is used to dealing with worker's compensation – a doctor who will say exactly what the insurance companies want him to say – and let somebody else – my internist – be the one to have the responsibility of running a bone density test when I am out of the hospital and rehabilitating. A lot of people wonder why this type of test would not have been done while I was staying at the hospital. When you examine the reasoning behind the madness, you begin to understand why the bone density test was not run. If the hospital does not have knowledge of the existence of a bone density test, they don't have to worry about being sued for malpractice. When I realized what was happening, I grab the yellow pages and hire an attorney to argue on my behalf so I can get the benefits that I am entitled to. When I told my internist that the hospital did not run a bone density test, even he was surprised when I said that. Unfortunately for me, there was nothing that he could say or do on my behalf.

Basically, my osteopath would just tell me a lie right to my face about how I got injured. What was the point of a trial when the deck was stacked against me? If I was going to win, I was going to have to tell lies in the deposition. I simply had no choice. If they were going to lie to me about how I got injured, despite the fact that I knew that it really was a cover up for something, let's just level the playing field so then it is fair. I'll just tell a few lies about the facts and circumstances. What's good for the goose – good for the gander.

I was informed by my attorney that if the osteopath did not agree with the fact that my injury was related to the work that I was doing, that I would have to drop my claim. It was going to be a cold day in hell for that to happen. It didn't matter that my father saw how filled my car was with telephone directories, he could not even testify. The court would not take that as acceptable evidence. He was not a credible witness. The only people who could be considered credible witnesses were my internist, who already decided not to testify and my osteopath who clearly had no idea what kind of case he was dealing with. I was ready to take this all the way to a trial, if it would have come down to that. My attorney talked on the phone with my osteopath several times, and in his heart of hearts, my osteopath believed that my fracture was a result of something unrelated to my job duties on the loading dock on that hot summer day. My chance to give my side of the story would come when my attorney had a deposition set up, which involved giving information as it pertained to my case to the lawyers for the insurance company that denied my claim originally. This would be a situation where I had the chance to turn the tables on the dishonesty that was in the air already.

THE DEPOSITION

Due to the fact that my injury was out of the ordinary, and I was filing a worker's compensation claim, the insurance company wanted to find out what they were dealing with. One day, I got a notice in the mail from my attorney's office that a deposition had been set up at the office of the insurance company. This would be one of the few chances that I had to turn the tables in my favor. Following by example, I just waited for my hand to be dealt. With the deposition being held within the territory of the insurance company's office, you would expect that people were going to interrogate me. Instead, what I was finding was a vibe of friendliness coming from the person who was asking the questions pertaining to my injury.

Questions were asked with regard to how I fractured my hip in three places. I knew that there was a lot at stake – and I was prepared to tell a few "white" lies for my own financial gain. My whole game plan was to put the insurance company on edge and get them to think that I had more information than I really had. I didn't want them thinking that I was afraid of them. After all, I had the upper hand due to the fact that my own situation was unique like no other. If there was one thing that I had that worked in my favor, it was knowledge about what happened. Any question that they asked me, I didn't waver in my answer. I answered each question like with the attitude that I knew the answer. Just like a game of poker, the idea is to bluff so that the other side thinks that you are more powerful than you really are. And that was the only way that I was going to see the light at the end of the tunnel.

On the one hand, I did not want this case to go to court. Because then the insurance carrier would have called in my osteopath, who would probably have said that my injury was not related to the work that I was doing, which was lifting telephone directories. On the other hand, I knew that if I answered all of the questions the way they wanted me to answer them, I would never have gotten the settlement that I have today. The saying "nice guys finish last" is definitely applicable in this situation. Because guess who insurance companies tend to listen to? The people who bring in business to them.

I was sworn in and asked a series of questions. Questions, such as "Have I ever filed a worker's compensation claim before?" to "Why did I take the job to deliver telephone books?" The deposition lasted about forty-five minutes. After the deposition, which my attorney attended with me in case he had to run interference, he told me that if they offered a settlement, that it would be in my best interest to take the settlement. That would be my only chance to recover financially from this mess that I got myself into by taking this job. In all honesty, I have to say now in my heart of hearts, that this was no longer a mess. It would eventually prove to be considered job security.

The one question they asked towards the end of the deposition was whether there was anybody that I knew of that could say that my injury and my work were related to one another. I sort of figured it would not get to a court of law, so I threw caution to the winds, and lied, saying that my internist could draw the connection between the injury and the hip fracture. Even if it did get to a situation where it would have gone to court, which I knew was *not* going to happen, my internist made it clear that he would not testify because he did not consider himself to be an authority – so in that sense, I knew I could get away with telling a "white" lie.

In my vindication, however, Dr. Metzger did say right in my presence, that based on the fact that I had taken Tegretol for fifteen years, that surely the hip fracture was connected to the use of that medication. And that due to taking that medication, I unknowingly had been living with osteoporosis and the day that I took this job to deliver telephone directories was when I found out about the osteoporosis. This could have happened during any other time. It could have happened during the time that I was relocating all the way across the country. God just made sure that it happened when I was working so that I would not be penniless. Thankfully, it did – otherwise I would not have ended up with a reasonable insurance settlement.

Because I had been living with secondary osteoporosis for about fifteen years, it would not be dishonest to say that the injury was related to the work that I was doing. One of the issues that the insurance company had was why I had waited two days to go to the emergency room to get treatment. I thought that I pulled a muscle, which explained why my system had about two bottles worth of pain medication by the time I got to the hospital.

A TRIP TO HOLLYWOOD

The inner struggles that my diagnosis gave me was not only the springboard for this book, but it was also a springboard for something even more rewarding. A trip to Hollywood, which would launch my freelance business off the ground, or so I thought. Because of my horrific accident, I decided to open my life up to the entire world and not only write this book, but also figure out how I was going to write a movie script, which involves completely different language.

If you have got this far into my book and didn't think that my life was going to undergo a radical transformation, guess again. There are some people that are resistant to change because they are afraid of embracing it. Because to accept change, means that you also have to embrace change. Some people can't live that way. They like the idea of security. They like to have routines that they stick to, day in and day out. They like to get up at the same time each day, do the same tasks at work each day, have dinner at the same time each day, go to bed at the same time each night. They like to get up and do their boring tai chi exercises every morning – and look like an embarrassment to the neighborhood while they are doing all this. Does that benefit society? Routines are boring for those with a message to get out to the world. Routines were okay – until I took the one job that ended up changing my life. For those out there that can't stand the fact that I have embraced change, my suggestion is to get over it – quick – because you know damn well who you are.

This was the one job that would not only change my life, but it would also benefit a lot of people who could learn from my experience. I felt that if I had been through this experience, and did not use the information and research that I came across as I was writing out my story, I would be doing a grave disservice to millions of people who are afflicted with the same set of circumstances – and don't even know it.

My father thinks that I should just get a regular job like everyone else. Good luck. I have tried with no luck to get a job. While our country is in a financial death grip by President George Bush, I don't think that I will have a job anytime soon. Not trained to think outside the box, she simply cannot understand why I have taken things to this length. I thought my mother would be somebody who would be there for me. I was completely mistaken. Something is going on in that camp, and I don't even want to know what it is.

After that, I called my entertainment attorney and had him fire a depiction release, which basically says that he agrees not to sue me for defamation or interfere with my projects – which is actually a sound thing for me to do, after reading that there exists a chance after my movie is released (if at all) that, out of nowhere, somebody will try and sue me for many reasons, one of which has to do with the fact that they want to try and claim that I stole their idea. I am not being paranoid. This weird stuff happens all the time in Hollywood. Until you have written a movie, you might not understand the reason for errors and omissions insurance. With our country being lawsuit crazy, just to get a bank to finance a picture, the appropriate releases have to be obtained from people who have an expectation of privacy in their everyday life. I basically offered $1,000 for him to sign the form and leave me the hell alone. He never did return the form to my attorney's office.

Since at the time that this book is being written, I also have a movie in the process of being green-lighted about my life experience, once it is green-lighted, I was looking at a very high six figure to low seven figure amount for my services. I had to find a way to protect my financial interests, and I was not about to let anybody get in the way of my dreams. He claimed that I was being unrealistic with my expectations. How do you expect me to get back to normal after my life almost ends? I don't think that it is that easy. As far as I am aware, he did not sign the depiction release agreement – even after I offered him $1,000 to just shut up and go about his business. As I have said earlier, those that try and sue me are going to be facing my insurance carrier – or the producer's insurance carrier – which had me added on as being an additional insured on their policy.

SLIPPERY FEET

After I was given clearance by the orthopedic to walk on my leg, I found out just how brittle my bones were. Walking around at times, it felt like my left shoe was going to come off my foot. I look down at my foot – and notice that my shoelaces are not loose. I walk around more on this foot. Periodically, my left heel, which already was determined to be a weak spot, according to the bone density and bone scan, seemed to not be supported by my heel. I was walking, but at the same time, I was limping around. A lot of the time, however, I would experience intense pain in my left foot. Sometimes, however, it was as though something didn't feel right when I walked.

When this happened, I had the life scared out of me. What if my heel fractured and caused some internal bleeding? For two years, I had not been rollerblading. As much as I wanted to get back on my inline skates, I had to wait until I had another bone density test to make sure that there was not a chance that I would end up with a fracture. All those years during my childhood that I periodically had pain in my heel and had to wear a heel cup now is explained in a different way. My skeletal system had been slowly starting to form holes in the bones, which would result in early osteoporosis.

I believe that everybody has in them the ability to write and tell a story, and given the right set of circumstances, this happens. Somehow, I realized when I moved away from my hometown, that things were going to radically change for me. I never expected to be working in a completely different field than what I graduated in, however. I was going to be completely different from anybody else out there, and I was going to be sure that everybody in this entire world was made aware of that fact. When this book started, it was just a book about what I went through. It was suggested to me that I turn what I went through into a movie. Hmm – good idea. There was only one problem. I had to be the one to write the screenplay. Ho w the *hell* am I going to do that? I had purchased a book on screenwriting, as I am always trying to implement new strategies into my overall marketing plan. It was suggested by many people who I told that I was writing a book about my experience, that I should write a movie.

I had a book to write. I thought that they were completely nuts.

Chapter 5

The Aftermath: Development Hell

I am sitting at my computer, my mind going in and out of consciousness. I really want to go to sleep. The fact that I am trying to make a sale is what really keeps me going, even though I have been working for at least seven months without a salary. I get up and go to the refrigerator and retrieve another Dr. Pepper. Figuring out how to re-arrange certain parts of my book to different areas, I notice out of my peripheral vision that there is a huge ball of fur sitting on the floor. Looking closer at it, Cozy, my cat is snuggled up next to my feet as I am typing this chapter. She probably just wishes that I would go to bed. I am sure that she would follow me as I slip under the covers. That isn't going to be happening anytime soon. Spending almost every night staring out the window as the dark sky lights up, my sleeping patterns have definitely changed over the past few months while I try to break through, somehow, and get my story told the right way – the way that Hollywood wants me to tell it.

I look at the clock at the bottom of my computer control panel. It is 4:35 in the morning – a time that I normally would be sleeping. I have my music playing to keep me from going insane. With no weather fronts, I don't have to worry too much about seizures, but I am getting eight hours of sleep – just not at the same time I would normally get it. I pick Cozy up off the floor, out of boredom, and start to get a laugh out of her. As I do this, she meows, surprised at being handled at this hour. I hear a low-volume growl coming out of her. I try to comfort her – then the unexpected happens. She jumps, grabbing my head for dear life with both of her front claws and tries to get a bite out of me, while growling like a tiger. It is unbelievable how the smallest creature can produce the loudest sounding growl...I am surprised she didn't try and fling my glasses across the room like she did a few weeks ago...

As I put her down and look out the window, the sky is lighting up as the sun starts to rise. My eyes start to burn. Throughout the day, I take a break and rest on my recliner. My chair of comfort, I would retreat there anytime I was deep in thought about something, or tired. More than half the time, I ended up falling asleep in the recliner for long periods of time. Gee, my bedroom is only ten feet away...

It finally occurs to me that I have a monopoly. Just like Microsoft, I am the only person on record who has epilepsy and osteoporosis. Using this as a weapon, I re-establish my priorities in life. Taking time off from school, I pray to God that I finally get somebody to understand what I had been going through.

As I had lost my health insurance after being cancelled due to more than $50,000 in medical expenses, I had to find a way to get a normal life back in my hands. Since I

THE AFTERMATH

After realizing that I was going to be on crutches for at least five months, and the fact that no doctor had ever seen a case like mine before, something had to give. It only was a matter of time before I started to write about the experiences that I was going through, and the struggles that ensued. It got to a point where I could not be emotionally invested in going to a day job. The economy just made it easier for me to do that. The good thing about having a president in office that thinks he is like royalty is what gives me the perfect excuse not to look for work.

Also, the fact that I live in the fifth largest city in the United States, there are more people than jobs. With a good number of people out of work, and more people than there are jobs available, I knew that I had something that nobody could take from me, which was life experience. As I realized that having an education is the same as not having an education, I dropped out of college for personal reasons. The aftermath of my diagnosis provided a question, which is, "Do I really need to have a college education at this point in my life?" As you read through this chapter, you may understand my line of thinking about why I say this.

I had to be careful with how I lifted things, especially the three gallon water bottles that I use for my water cooler. My internist put me under strict orders not to lift anything more than twenty-five pounds. The real aftermath, however started when I left my hometown, a crippling relationship, and joined The Real World cast. What would follow would show the types of things that distance can do to those when they are separated from one another.

WELCOME TO THE REAL WORLD

"Seven strangers picked to live in a house, live together, work together, and have their lives taped. Find out what happens when people stop being polite – and start getting real. *The Real World."* I will be the first person to admit that I am addicted to MTV's *Real World* series. Twice a year, the show is set in a new city and different cast members are picked. You know, that show where they stick seven people in a house and they have to live together for several months. The reason I watch that, partly, is because I notice what happens when one person decides to pick up their roots and live somewhere else for a period of time – or in my case permanently. Anybody who watches that series on a regular basis knows that nobody who leaves their hometown is the same person coming back to their hometown. It is the perfect way to get you to understand this chapter and what I am going through on an intellectual level.

When I watch that show, anybody who comes onto that series with a relationship back "home", in almost all cases they are no longer together after the show is over because of the change that one person goes through relative to what is going on from where the person came from. Sometimes you have the occasional cast member who is constantly talking on the phone with their significant other. Actually there is usually at least one of those each season that does this. What was the point of even getting on the show in the first place? I watch that happen, and I think, "What are these people, antisocial?" I only say this, because my parents are angry at the fact that I have been out of communication

There is a certain comfort level in sticking with what you know and not throwing caution to the winds. I have been living in Arizona for three years now, and by the time you are reading this, assuming that I have received a contract from a literary agency, I will probably have moved again since my stomping ground gets worn out every few years. Just throw about a half million dollars or more in my direction, and that will be enough to make me pack up and get the hell out of here before the cavalry comes calling.

That is why I have made it my policy to limit my communication with my family strictly to email – which my mother is not happy with. Sometimes, she won't respond to my email because she wants to manipulate me by forcing me to call. The phone never rings, because I am not dumb enough to take the bait. I don't play based on *other* people's rules. People play based on *my* rules. I don't like the idea of someone being able to string me along and play with my emotions. There is a good saying that says, "If you give an inch, they'll take a mile." That is the absolute truth. The only people who I do call are people who I have met outside of family, who are supportive of my visions, and can give me the necessary space that I need so I can go about my life, and don't nag me about why I never call.

Family members can sometimes be considered "toxic friends". I started to notice how I was resenting my family – especially when I financed a car. That was the beginning of the change that would result in the freedom that I have today. I notice the coldness that I experience, especially from my mother, who does not seem very happy for me. I know that when this is in print, I am going to get killed for saying it, but I really don't care. Anytime I go home, there is an uncomfortable silence that words simply cannot describe, especially when I am sitting in the car for thirty minutes after leaving the airport. It takes a real understanding person to see why a person does what they do. But then, if you really thought that you understood me that well, you wouldn't be reading this chapter.

When I told my mother that I was planning on relocating to Arizona, she was completely devastated by that, like I was tearing the family apart, almost accusing me of being a traitor. That was the eventual nail in the casket, which would serve only to push me very far away. My father was asking me why I couldn't just get another apartment in Milwaukee. Before I moved to Arizona, I didn't know how much was going to change over a one to two year time span. My mother was in angst because I dropped all communications with them, which just served to push me further away when I was home during my ten year high school reunion in 2004. She asked me on a regular basis if I ever thought about returning home. Oh, never mind that six months after moving to Arizona, I almost died.

Since she has been supporting me financially, my mother resorts to underhanded blackmail, which I think is an attempt to make me move back home. All that accomplishes is to push me further and further away. Keep up the good work. Maybe one of these days, you'll get a clue, mom. One of her battle tactics is to blackmail me. She will say "if you want to stay where you are, you have to get a job." Another thing that she tends to emphasize is how my work will never sell. Yes it will. Anybody who is a literary agent and sees how my story is going to sell like hotcakes is going to do anything to get me to relinquish the copyright to my work and license it to them for publication, assuming that they pay me a fair amount of consideration for that privilege.

What my mother is really saying is that she is unable to relate to my projects on any given level and wants me to give it up. If I were to say that to her, she would eat her words and say that she is supportive of me. Well, part of the reason I focus so much on my projects is because the phone isn't ringing, even when I do apply for jobs. I am not going to apply for jobs ad nauseam. I think on a realistic level and put my resources where they are most likely to be of benefit down the line. Maybe not immediately, but months down the line.

Never mind the fact that I live in a city about the size of Los Angeles that shows no sign of getting smaller in population size. In one of her recent emails, she said "that if my brother could find a job in a year, that there is no reason why I can't." Be realistic. Milwaukee is the 19th largest city in the United States. When I lived in Milwaukee, I could get a job *easily*. Phoenix used to be the 6th largest city in the country when I moved here in 2001, and now has become the 5th largest in population size. Add all the snowbirds that tend to flock to Arizona when it is winter in the northern part of the country and we could easily be the fourth largest in size. With three million people all looking for work at a given time, it makes it a lot more difficult to find steady employment. Do you see my point?

She also seems to forget that certain experiences permanently change the social fabric within a person. I am not saying this to be an ass. I am saying it because it is the complete truth. Anytime she would call me, it wouldn't be to ask what I was doing or show any interest whatsoever in what I was doing, but it served more to tell me about what *she* was doing. I don't need to know what she is doing. She goes to work, comes home, and stays in a routine. What in the hell is so damn glamorous about that lifestyle?! My immediate response is to hit the delete button on the voice mail. Some of us are busy and can't afford to be wasting time with those of us who are not in any way understandable of our situation.

I know that some of you are going to say that I should show interest in what is happening at home. But, I am the one who took that leap of faith, moving two thousand miles across the country, and then made a discovery about myself that I would not have found out about otherwise, that in all actuality, saved a lot of people from a big hospital bill. I probably would not have taken that big leap of faith, if it had not been for my next door neighbor moving to Tucson. The rest of my family has not changed at all whatsoever. The world is real big for a reason. The reason, is that it is meant to be explored. There is an element of truth to the saying "No man is an island."

When I went home, my mother went to all lengths to suffocate me, not providing me with the necessary space that I need. Then I came upon a little idea, which I conceived just to piss her off. I invited a friend of mine that lives in Duluth to, in essence, keep me company for the next two weeks and turn the house upside down. That answered my questions as to what the ulterior motives happened to be. She wanted me all to herself and not to be in contact with anybody else. Well, now you know why I never call home, unlike some of the cast members on *The Real World* who have lost touch with reality.

After a while, you can understand why a person who is selected for one of these *Real World* series is happy to meet new people. A family that does not socialize outside of themselves gets to be dull and boring. And why don't I call home any more than once every six months? If I wanted to call home on a regular basis, I wouldn't have moved from home. Case closed. End of story. And I am not planning on moving back home for any reason whatsoever. For those of you out there listening to this, start taking notes on that last sentence. You know *exactly* who you are.

CAN OF WORMS

At this point in my venture, there were people who were preying on me like vultures. They couldn't have cared less about me. They only cared about one thing – money. And I was going into this venture completely unprepared for what I was about to deal with. I never had to deal with things like the paparazzi or people who would have a mob mentality or stalkers, but there were hints that I was getting that I should be putting some additional security measures in place – as well as getting the hell out of my apartment the first chance I had and moving to more secure quarters. I didn't know about this until I found out that my dad had manipulated the information with regard to my settlement out of my attorney. I sensed some ulterior motives – and put that money in a secure place. He went as far as to say to me on the phone *"Eliot, I talked with your attorney and he said that you are getting $10,000 dollars!"* Saying this to intimidate me, I did what I needed to do. Nothing.

At that point, I didn't have a clue about what settlement that I would be getting, so I am not going to comment and end up taking the bait. As I was able to detect some jealous flair in the tone of his voice, I did what I figured was appropriate. As I would find out, he was not the most supportive of me when I was trying to get my projects in the right place, regardless of what my mother says. What I would find out, is that my parents would lie about how they supported my visions in an effort to avoid looking bad. My mother went as far as to say that "my work might not sell." Who was she to make this comment, even if it would by some chance turn out to be true? She had not even done the research like I have, so how could she possibly have the right to say that? My screenplay will sell – even if I have to risk my health and well-being to make sure that it sells. I already have picked up some signs that Leonardo DiCaprio may have contacted my attorney to express an interest in playing the lead role.

SHIFTED PRIORITIES

I didn't know what the hell I was signing up for. I had no idea that my entire lifestyle was going to be changing. I am sure that the word got around pretty quickly around Wisconsin and some people tried to track me down. The one thing I did that made my life a little bit easier to handle – I disconnected my landline phone only two months after I was discharged from the hospital – not realizing that I was keeping myself from going insane when development hell came about and just about everybody and their brother knew about my story in one way or another because of my website, which my father told me to tear down. Well, if I had done that, I don't know what I would be doing with my time. Maybe somebody doesn't want the truth to be let out. [1]

THE KEY

I have to say, that had my manuscript that I originally wrote never been seized by the sheriff's department of San Angelo, Texas, for the purpose of using that as evidence, you would probably not even recognize my name. Because things worked out the way they did, it worked as a way to make something come out of this for everybody. My book would have been just like any other in the bookstore. Because my name was so widely known at this point, it was not necessary for me to hire a publicist – I had my own form

---

[1]

of publicity. The word about my story was spreading like the wildfires in southern California. I was watching this all come down on the internet as I did my surveillance. There would be times when over a two day period, I would have forty or fifty hits to my website.

It was actually my personal foundation that told me the advances that are paid to authors are not substantial. This resulted in an entirely new business plan. I had purchased a book on screenwriting, out of curiosity. It was the movie that would possibly save my book from the trash piles. Almost all the agents I queried said that I lacked the authority to write a book on my experience because I did not have a medical degree. Gee, if the doctors don't know much about the link between epilepsy and osteoporosis, where am I supposed to turn to? And who really does have that flair of authority? Is that to say that despite being through a situation that almost kills me, that I have no business telling people how epilepsy is related to osteoporosis? I hardly think so!

Since the movie would be coming out before the book could even be released, that would be a major windfall. There was no way I could afford to self-publish my autobiography before getting some form of an option payment from a producer. And at this time, no agent was going to touch me after they found out that they would be dealing with "hot" material – which is why I re-wrote my book. I didn't like the way my book originally was, anyways. It was really the working material, or the prototype, for the material that I have added to the current version of my book.

It was mentioned by a few people that I told my story to that I should try and turn my entire story into a movie. I shrugged that off, as at the time, I was trying to get the rights sold to a book that I was writing on the link between epilepsy and osteoporosis. After slowly finding out that the agent I was dealing with was nothing but a shark, that is when I embraced the idea of writing my life story for the silver screen. Oh, shit. That also means that I have to learn how to write the darn movie script.

Tell this to somebody who has not written a screenplay in their entire life and is not used to writing with any specific style. Yeah – that went over really well with me. Not only could I not write the character's inner thoughts, but I would have to learn how to package the entire script the way Hollywood wanted it – using the proper font, proper margins, certain words capitalized – specifically words which indicate action or noise, and I had to learn how to write all of this in what is referred to in Hollywood as a master scene[1].

HOLLYWOOD – AN EXCLUSIVE CLUB

You are rolling your eyes wondering, why should it matter how a script is packaged? It matters because Hollywood has grown used to having things a certain way and it isn't going to be changing in my lifetime. Another good reason is the cost for producing a movie. Studios invest hundreds of millions of dollars that go into production of a movie, and they don't want to go over the budget.

The rule in Hollywood is that a properly formatted script page is the equivalent of one minute on the movie screen. Say for a reasonably budgeted movie script that costs ten million dollars, that means that for a script that is 120 pages, each page is going to cost the movie studio approximately *$80,000 per page* to produce. There is a reason that studios tend to reject 98 percent of everything that comes through the studio, and why anything that is high concept seems to get the pink slip.

I remember the first time I submitted my movie script to a script coverage company to be read, I was told that it was not formatted according to acceptable standards. I had the script double-spaced (it was supposed to be single-spaced), I had things in the wrong place, I had words capitalized that weren't supposed to be capitalized, and I had been using Times Roman as my font, instead of `New Courier`[2], the appropriate font, which is much easier on the eyes of readers who read many scripts every day. If you look at the way the font is designed, it is very similar to the font that you would find on a typewriter and is much easier to read. That is why there is such a fuss made about the font that is used for movie scripts. To address these issues, I checked out *The Screenwriter's Bible* from the library, which showed the appropriate way in which I should put the script together. Somebody actually had the human decency to put these rules in print.

Anything that is submitted to movie agents has to be formatted properly. I had a book that I bought on screenwriting, since I was wondering how all these movies get

---

[1] All scripts are written with master scenes that specify a location where the story is supposed to take place.

[2] New Courier 12 point font is the industry standard font used on Hollywood movie scripts because it is easier on the eyes of those who read as many as thirty or forty scripts every day.

made. Since I was waiting around for something to happen with my previous manuscript – the one that my ex-agent was royally screwing up for me and not telling me about the fact that she was screwing it up – I started to write a screenplay off of my experiences. I had nothing better to do. It was better than sitting around doing nothing but finding new ways to annoy Cozy, the twenty-five pound hell-raiser who liked to tear from room to room, which made me go nuts, leaving behind papers that just flew all over the place. And the cat blames me for why my apartment is in a mess as though it was jut hit by a cyclone.

Readers in Hollywood are used to seeing screenplays in a certain way and the last thing I wanted to have mine end up at the bottom of a scrap heap. It got to a point where I had to educate myself to learn a system that had been in place well before I was born and had not changed since its inception. The movie studios came into play around the time the stock market crashed in the 1940's. I was not in a mood to learn how to do things a certain way. However, just to save myself, it was something that I was going to have to learn.

Based on the fact that I had been through an unusual experience, it was worth putting myself through this torment. And this was the most rewarding part of all. I owed it to myself – and my mother – to make a lot of money from my life story. Talking with Joey Sayson, my entertainment attorney, who represents my screenplay and looked after my creative rights, he said that I was eligible to get life story rights in addition to the actual script sale. Not only that, he would be instrumental in trying to get me a role either as a creative consultant on the set, and/or an executive producer credit on my screenplay, which would not only bring in more money to my bank account, but would completely change my life as I knew it. What does that say about the idea of getting a college education? I finally had something that nobody else could take from me. There are those that still don't understand why I have chosen to drop out of school and pursue other ideals that are more creative. My diagnosis and the creation of my screenplay is exactly like Bill Gates and Microsoft Windows.

The movie also might have added value to my autobiography, which probably would have gotten a meager advance as a first time author from an agent to someone with no track record as an author. My life sitting in my apartment in front of my computer all day for hours at a time was wearing thin. I wanted to expand my social network somehow. It would be interesting to actually meet the actors who would be playing the different roles for my movie. I already had an idea of the person who should play the lead role for my character – Leonardo DiCaprio – after watching his role in the movie *Catch Me If You Can* where he played the role of Frank Abagnale Jr., who, in the process of trying to run away, ended up learning how to kite checks and move from place to place, ducking the law. It was by doing this, that his character starting working in the FBI Financial Crimes Division and helped to design security enhancements for company checks. Being the only one who knew about this earned him millions of dollars a year. Likewise, I held a monopoly and am trying to use the unique nature of my disease to put together a movie in the hopes of making a seven figure salary.

PREDITORS

In my introduction to this book, I mention that this book almost did not happen. I ended up getting burned by a dishonest literary agent, who I will not name because I don't want to get sued and (ahem!) I eventually will have to work in this town. Let me start the assault by saying that I was fed nothing but lies about my abilities, about the marketability of my work, and that my work was even seen. She also neglected to tell me that she was under investigation by authorities for bad business practices. This agent had changed the name of her company many times to escape authorities. In the end, she was charged with "theft by deception and mail fraud." Perhaps you recognize who this agent might be, but I will not name who it is because as of the printing of this book, there is no public record of her even being caught and tried for the crime she was charged with, and I don't feel like getting sued for libel since there is no public record of her ever being caught for her crimes.

I ended up paying this dirty scoundrel $300, which I thought was going toward getting my book published. Because of the fact that I had the money, I had no problem questioning the amount, which was nothing compared to the settlement that I received as a result of my hip fracture. The check that I wrote was like a drop in the bucket. The bad thing, is this investigation delayed my book, but the good news, is that it may have scored some sympathy points with the production companies that started reviewing my work in the summer of 2004. No pain, no gain.

I don't know what my agent was doing with all of this money, but there is a long song and dance to what happened. I was on the phone with her, in the process of putting my movie script together, and I was told to go ahead and send it. Alarm bells started going off in my head. She had not even sold the proposal to my book, and she is telling me to send the movie script to her office? As inexperienced as I was dealing with the publishing world, something told me that this was not a good idea – and I listened to that voice in my head.

There was a voice in the back of my head telling me not to send the screenplay to her office. Somehow, I had a feeling that I was getting the royal screw job – and I would have had that happen if I went against the voice in my head. I asked if there was any catch, like would I have to pay additional money for sending this to her. She said that I would, since this was a separate contract. I decided that if my book sold, then I would send the screenplay to her. That never ended up happening – because only two months later, she ended up having a search warrant served on her home and business.

GUILDS

As I was doing research on how to get my movie script into the "right" hands, I found that producers will not read any work that is submitted by agents that are not signatory to The Writer's Guild[1]. The Writer's Guild is divided into two parts. The dividing line is the Mississippi River. Those who lives west of the Mississippi River is governed by The Writer's Guild of America, West. People who live east of the Mississippi are governed by The Writer's Guild of America, East.

This organization is similar to the counterpart that deals with publishing of books, The Association of Author's Representatives. The primary difference is that both writers and producers can maintain membership in The Writer's Guild while only agents are members of The Association of Author's Representatives. I look up my agent's name on that list – she is not on there. If nobody will read work from a non-signatory agent, think about how many publishers read my initial manuscript. I would guess none of them did. That should tell you what would have happened if I had sent my screenplay to her

---

[1] The Writer's Guild is a union that represents writers that work in features and television series. They specify a minimum that a writer has to be paid for the sale of their script. Producers who are signatory to the guild have to follow the guidelines outlined by The Writer's Guild.

attention. No producer would be able to read it, because part of the working rules of The Writer's Guild, which most producers are members of, is that the producer who is signatory to The Writer's Guild is not allowed to read work that is submitted by an agent not affiliated with the Guild. The only exception to this rule is with first time screenwriters, who are not yet members of The Writer's Guild. A producer can read work submitted by a writer with non-signatory status, under the condition that upon selling a spec screenplay or accepting employment, they become members of The Guild.

My first book never sold, mainly because this agent is nothing but a ditz. She is the type of agent who couldn't get a project sold if somebody paid her the entire mint. Somehow, I started to pick up a vibe that she may not have been telling me the truth when I signed with her. It was this incident that started me on a journey to find out the real truth. My friend, Lynn, who got her book *Holistic Parenting* published, was expressing concern about this agent around the time that I was at her house for Thanksgiving.

At a Thanksgiving party, Lynn expressed concern to me based on her knowledge of the publishing industry. What I never told her, was that I decided to get behind the wheel of my car and drive about one thousand miles all the way into the center of Texas to do an investigation on the physical address where my stuff was being mailed. It was what I found out that made me think twice before I sent her any more of my work – not to mention made my stomach do knots. The sight was not pretty. I don't know where my mail was really getting sent, but it certainly was something that I started to get concerned about.

THE ROAD TRIP TO THE TRUTH

San Angelo, Texas – May 2003 – Six months before I started writing my movie script, I had some weird feelings about where my manuscript was sitting – as well as a lot of free time since I was in between job assignments for my temporary agency. At first, I was hesitant to do this, because it would mean two long days behind the wheel with my eyes burning up. On the other hand, if I didn't find out for myself, this book would probably never be released and into your hands like it should be. I took it upon myself to take a joyride – all of 1,000 miles – out to Texas to do a check on the physical address where all of my mail was supposedly going. I wanted to see if I could pick up any indication of what this inner voice was telling me. Speeding all the way across the state of New Mexico and into the center of Texas, I was kept awake by drinking a huge amount of Red Fusion and Dr. Pepper, sodas which have more caffeine in them than the strongest cup of coffee.

Following the directions that I had printed out on Mapquest the day before, I arrive at the destination, and see nothing except ramshackle houses. I look around, and there is just empty land with nothing developed on it. Am I on the right street? Looking at my map, it says to go one block and then about halfway down, I should find something with that address on it.

If she is doing so well, then why in the devil would she not have a separate office? I knew that something was not being told to me. I was in disbelief – how could somebody tell lies to me over the phone? This was a bad sign. As I reviewed my notes on my map, the directions led me to an area where the only building was a ramshackle shed or farmhouse. I was on the right road with the wrong timing. I turned around and made my way back to Phoenix. What would follow would shortly thereafter would be a major decline in visits to my website, which I took down in December. Upon re-establishing the site in February of 2004, I was inundated with hits – only to have the visits drop off again after, with no other option, I had to inform the visitors to my website that the literary agent that I thought was working for me had a search warrant served on her home and business in late January 2004.

I found out about this bullshit in March – and that was when the hits to my website went south. At the time, I was toying with making a movie out of this experience after listening to some words of wisdom from some people that I was working for on a temporary basis only a few months earlier. Only I happened to know that there was some twist to why the business was abandoned. Currently, she has charges of theft by deception and mail fraud pending against her, but has not yet been brought to justice since at the time the search warrant was served on her, she disappeared into nonexistence. That would make sense, seeing that the address to where my mail should have been going was ending up at an abandoned house that probably was not livable. It has been almost a year and the detective that I gave my information to has not gotten back to me. All the times that I was sending mail to the address that I was staring at on the page that I printed out on Mapquest, I had no idea where it was really going. Because on the road I was on, there was not much of anything there. Below is a chart that shows the visits each month to my website and where the hits dropped significantly, as I had to print, with extremely large letters on my website page that had my book on it, that due to my literary agent being investigated, my manuscript had been seized by the sheriff's department and was being used as evidence.

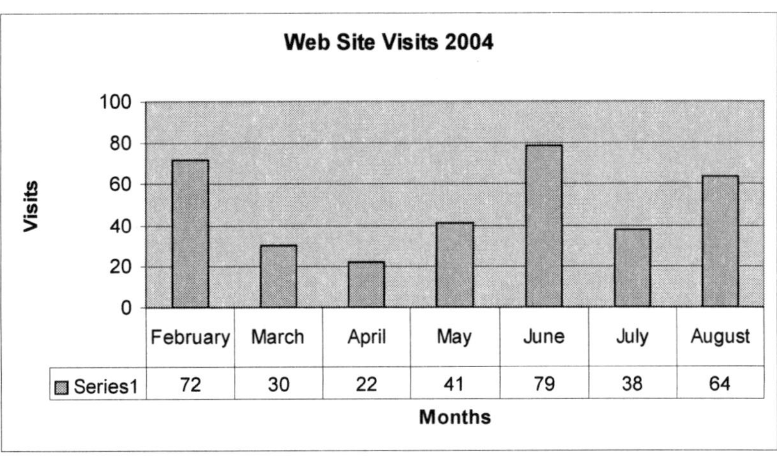

	February	March	April	May	June	July	August
Series1	72	30	22	41	79	38	64

As a result of the information that I had to post on my website, to be honest to the people visiting my website, my visits as few as they were took a major nosedive upon my discovery of what was going on in March. I was almost thinking that life for my freelance business was coming to a screeching halt. That was made me throw the last of my resources to where they needed to be, in early March, which was in California, where I did a one-on-one script consultation with a producer who regularly meets writers to give them advice for what they need to do with regard to improving their spec script. At the time, I was also unemployed and borrowing money from my parents due to hard times. Over the long haul, the web site visits returned to the same level as they were in February after I got the information that I needed to get my movie script changed in an appropriate manner.

SHIFTING FOCUS

I kept adding to my book and reading the books that I was buying as it regards to entertainment and publicity and about screenwriting in general. My script originated starting in November 2003, when I had this conversation with her. But, based on the clues that I picked up when I drove out to Texas, I decided against sending her the script, which would have been the death knell for my business and my reputation. I was having no luck whatsoever, tried to find an agent in California. It is an absolute catch-22. No agent wants to represent new writers, but not many production companies or publishers will look at work that is submitted by writers who are not represented by an agent or a manager. The whole process has me frustrated. Until I got the attention of a production company, I ultimately shelved the autobiography that I had written with the intention of either self-publishing my book or getting a referral to a literary agent by my manager when I signed the papers with regard to my movie. As economic times would mount, that left me no choice but to self-publish my book initially until it would be picked up by a trade publishing house.

This agency is not the only one out there pulling this trick, either. I had another contract come to me via email, and I immediately started looking for a catch. They wanted $115 for what they call "marketing" costs. I immediately trashed the email without giving it a second thought. Take something from my experience. If an agent asks for money upfront, immediately stop your association with them. I don't care how nice they are or what they claim they will do for you. They don't know how to do their damn job. People that ask for money upfront are not agents. These agents often use praise as a way to get people to drop their guard and take the bait. Agents deduct their fee when a sale is made on your material, and not before. No exceptions. Here is another similar letter to the one that my first agent sent. It is from a company that I was going to send my screenplay to. The first thing they did was ask me to give them my address so they could send me "instructions" on how to submit my story. Here is what the letter they sent to me in the mail said:

Dear Mr. Trimberger:

In order to offer you representation, we must determine if the manuscript is salable. Please forward your material together with a self-addressed stamped envelope *and the evaluation fee of $395 to [company name deleted]. If the manuscript is not finished or is in proposal form and you wish to have us look at it, the fee is $5.00 per page.*

The evaluation will take 3 weeks from the date of receipt. We'll get back to you with a letter agreement that will constitute an acceptance for representation by this literary agency or a letter declining representation with comments and criticism.

If we are successful in selling your manuscript, we will refund the fee out of our commission. We look forward to hearing from you.

Another trick that some unscrupulous agents use is by saying "What you submitted is good, but it needs work. We recommend the following person to help you" and get you a referral to a book doctor. What they don't tell you is that they might receive kickbacks from this book doctor. That almost happened to me, except I had the nerve to turn the tables on the person doing this to me. Knowing full well that I wanted to make sure that this didn't happen a second time, I made sure that whoever took my work was not planning on charging me upfront. It was about to happen…again. I questioned the company's status with The Writer's Guild, which they said they were a member of. I had checked the online database just hours before they claimed this, so I thought that perhaps the guild did not have their database updated with the right information. So, I picked up my phone and let my fingers do the walking.

I called The Writer's Guild's offices in California and checked to see if this company was in the computer system like they were claiming. The person who was on the phone said they didn't recognize the company as being a guild signatory. Below is a copy of the email correspondence I got from one of these companies that I essentially turned the table on by calling their bluff:

"Hello: Our literary agency has charged a reading fee for over 22 years and it is not unethical. We deduct the fee from the sale when it is made. If you do not feel comfortable, don't do it. We are in the contract 2004 as a signatory to The Writer's Guild of America, West. Your sources are unreliable."

What this amateur does not know, is that I called The Writer's Guild to verify whether or not his company was signatory. Whomever told me this line of bullshit probably thought that I would be stupid enough not to pick up the phone and call the Writer's Guild to verify those facts – especially when the agency was not covered under the Guild's collective bargaining agreement. Do people think that I am stupid? That I am not going to pick up my phone and confirm what they claim to me? After the line of bullshit that my ex-agent who had my book was giving me, I background checked EVERYBODY. There was a period of time that I simply went without anybody on my side because, as I found out, these companies are about a dime a dozen. I was lucky to be able to find an attorney that would agree to meet with me without concern for what relationships I had.

The people in the guild are not going to lie to me. After my experience with my ex-literary agent, I knew when he said that "we deduct the fee from the sale when it is made", he had to be up to no good. What this really means, is "We are not good at being agents. So to cover our butt, we have to charge you upfront an arm and a leg." This was the same language used by my ex-agent. A real literary agent *never* should ask you for money upfront before a sale is made. The fact that they had to ask for money upfront was like saying they didn't know how to be an agent. That is the number one deal breaker. People who are in the habit of playing these con games probably tell themselves that what they are doing is not unethical for any number of reasons. When he said that I had to pay to have my screenplay evaluated, that was the red flag to trash that contact.

Once I am a member of The Writer's Guild, I can't work for agents or producers who aren't signatory to the guild, so it is my responsibility to know that information. There is a code of ethics that I am going to be expected to uphold. I read upon getting my production contract, which qualifies me immediately for The Writer's Guild, that I am required to immediately make a $2,500 donation to The Writer's Guild so that I can continue to be employed. If I don't do that, I have to stay independent and not have my screenplay produced. Part of the working rules of the guild, is that I cannot be sending my manuscript to any producers who are not covered by The Writer's Guild. If I don't verify that, I am in violation of The Writer's Guild working rules. No producer would have read my script, since it would have been coming from an agency that was a non-signatory.

Another thing this one company did, was ask initially for my mailing address so they could send me instructions on how to submit my story. I really should have seen this one coming, big time. What they were going to do, was tell me where to send $395 to read my script, which would have done nothing except grow interest in another person's bank account. Gee, I've been down this road once before. Reading is an agent's job. I shouldn't have to pay for that. Why don't they tell me in email the physical address as to where I can send a copy of my script if my story is so fascinating? That is what a legitimate company does. A real company also makes you sign a release form, which is really to protect the writer with the fact that there could be similar ideas in development around town. There are people who think that releases only protect the producer. They protect both parties, because when a lawsuit comes up, they generally will come after the producer and the writer, or whoever has a lot more money.

Then there is an example of a person who said that they couldn't put their name on my submission to a producer because my script was not written according to "professional standards". He was also trying to charge me $500 to do a "critique" of my screenplay. When he said that he was also a professor teaching screenwriting, that spelled "amateur" in my mind and I dropped contact with him. Treat your relationship with your agent like you would a real estate agent. If a real estate agent is only part time, that says something about their commitment to their clients. He was also not listed in The Guild's agency list and – you guessed it – he said that he was a member of The Writer's Guild.

If you can't locate an agent's name in The Writer's Guild database, then that spells trouble, even if they are claiming to be an agent associated with The Writer's Guild. Don't ever trust anybody, especially when it is your professional reputation that is being played with. If there is one thing I have learned it is not to take an agent's word about being in the guild. I cannot believe how many agents will outright lie to people like myself about being a member of The Writer's Guild when they most clearly are not signatory to the guild – and then have the nerve to believe that I am not going to pick up the phone and confirm these facts if I can't find their name as a guild signatory. This is different from the author's guild, which is for people that write books. The Writer's Guild protects the interests of screenwriters who work for studios and production companies.

I had a script coverage company, who also got themselves onto thin ice with me, who I was paying, in their first round of notes say that my screenplay was not formatted according to professional standards. So, I could understand if this agent made this comment at this point with regard to my screenplay not being formatted according to professional screenplay standards. Nonetheless, he still should not have even been bringing up the subject of money prior to a sale being made. I did more research on screenwriting, and the next time I submitted to this screen coverage service, their comments said that progress was made in formatting my screenplay in accordance with professional standards. Then, this ditz comes in out of nowhere and says that my screenplay is not formatted according to professional standards. This told me how much he really knew about screenwriting. Nothing. And I was not dumb enough to give him the money – but I did tell him in my own way that he was full of shit. I have no problem doing this. I will continue doing this as long as there are idiots that take me for a fool. "Fool me once shame on you. Fool me twice, shame on me."

I don't see anybody doing too much about the problem associated with literary agents that charge fees upfront. You would think that writers are protected from people like this. Unless you have been through this first-hand you would never understand the serious impact this can have on your financial well-being – not to mention your sanity, with your family breathing down your neck to start producing income. When I come up with a solution to the problem, I will certainly let you know. These people are about a dime a dozen.

When literary agents first came into the picture, they charged reading fees to evaluate your manuscript. As time went on, that was looked at as unethical. Now, there are some agents that are slick in the language that they stick into their contracts, which are about as valuable as the stock market while President Bush is running the country. Now, instead of reading fees, you are more likely to be charged "marketing" fees upfront by a slick agent. This is the very same thing as a fee that was charged to "read" your work, but instead of calling it that, we'll just say it is a "marketing" fee to call things fair. Isn't this all part of an agents job? Why are they even charging money upfront when it is clear that that is wrong? In the present day, some agents will refer you to a book doctor and receive a kickback from the book doctor. Why do they do this? By doing this, they can avoid getting caught and still be able to illegally collect money from starving writers. These are all schemes that are done by people who simply don't know how to effectively sell a manuscript without having to charge some sort of upfront fee.

What I am trying to point out, is that there are no licensing requirements for literary agents. They are about as unregulated as bounty hunters. Anybody can have business cards printed up, saying that they are a literary agent. Does that mean they are truly a literary agent? No! Credentials, credentials, credentials is what it comes down to. Check references. When I showed my former neighbor the letter she sent to me, she was indirectly expressing her concerns. She was knowledgeable about how the system was supposed to work, since she had just recently had a book published. I just wonder who it was that gave her that information when she needed it instead of making her find out the hard way that the agency that she was signed by was nothing but a complete ditz. In a twist of fate, it was this same company that was going to save me from the trenches in a way that I could not imagine.

RELATIONSHIPS

There is no other industry in this country, or the world, that relies as heavily on relationships like you would find in Hollywood. Hollywood is a business that is all about relationships, especially with the relationship that you have when you are managed by somebody. When I am assigned a manager, I will probably be talking to manager more than I talk to anybody else, like on a daily basis. Any job that you go and work at, you don't have relationships with people. There is a hierarchal chain of command. That is such an unhealthy way to make a living that I don't know how I was able to make it this far without going crazy. Nobody really gives a shit about anybody else. I am not yet on the entertainment circuit, but one thing is for certain. Once you make it over the hurdle, things become less complicated. There are still things that exist such as office politics and all that, but there is one main difference. In the entertainment industry, to a certain extent, people learn how to get along with one another instead of trying to one up another person. Making a movie is a collaborative process where you have to satisfy many people who are emotionally investing their energy in developing and producing the movie. To a certain extent, there is an element of give and take in Hollywood. Can I refuse to listen to people? Yes. But, I can still be replaced if I don't listen to the suggestions made by the people reading my script, especially once I cash the option check. Once I sell or option the material, they own the material and I have to make a lot of people happy. Everybody has their own creative input, even people who don't do that much writing.

You will never find anything of that nature working in a business setting. In a business setting, there are expectations to be met at any and all costs. You have quotas to meet. If you don't make enough sales, you are replaced by someone who can make the sales happen. You either meet those expectations, or you get fired. What a really nice system.

There were times I would be working for a temporary agency and they would put me on a "do not call" list because I for one reason or another, did not work out for them with one of their clients. There was a double standard. Your temporary agency can lie to you about anything it wants, but don't you dare lie to them. Any company you work for, there is something called employment at will, meaning that either party can terminate the relationship for any reason. Within the entertainment industry, however, under the guild's collective bargaining agreement, I am the one who has the power to terminate that relationship if ninety days go by and I have not found work through my manager or my agent. That is a protection given to screenwriters who get signed with someone who is unable to find writing assignments for them.

When I sign with my manager, which is a contract about as big as my entire script (120 pages approximately), I am signing for a specific period of time. I can sign a contract for whatever length of time I want. Usually, people sign for anywhere from a few months to several years. From what I have read in doing my research on the internet, some people choose to stay with the same manager throughout their entire career and then some switch when it is necessary. Then there are times that writers and their managers choose to go their separate ways because creative differences get in the way of the writing. By the way, there is a definite difference between managers and agents, which is the time that is devoted to the writer. Managers spend a lot more time with writers than agents do. Agents act more like salespeople, whose job is to sell material that is ready to be sold to the spec market. Managers are much more likely to take a new writer and elevate their careers. The agent will come later. That is what referrals are all about.

Just like agents, managers are not a regulated industry, and there is no ceiling to the percentage that they take upfront. If somebody signs agreeing to twenty percent, that is the amount as stipulated in the contract. I personally would not agree to anything higher than fifteen percent. Because then if you have an agent, that is another ten percent, my entertainment attorney gets about five percent, and then the IRS, who knows how much they get. I let my accountant deal with that. (I have an associate degree in accounting. I just have better things to do than being a bean counter.) Managers work for their client to guide their career. They also spend a lot more time with writers than agents do. Agents are only there as long as there is something to be sold. Managers are there for the purpose of growing talent. Managers can also sell screenplays, but they have to do that through the writer's attorney so that a fair agreement is reached.

The manager makes sure that my needs are met, such as getting a referral to a literary agent that won't screw with my head, a business manager, which is the same as an accountant, publicist, and other people who will make sure that my company thrives and prospers. This is an industry that is fueled by relationships. No other industry cares about the person as much as they care about the product. When something comes through with my name on it, they will surely remember it as something that must have value based on the last thing screenplay I have written. Hollywood is one of the few industries where you name has a true tangible value.

For example, I found out through something that I read on the internet, that Leonardo DiCaprio's attorney's allowed him to actually copyright the use of his name back in 1999, around the time that he starred in the movie *Titanic*.

That is why once my script sells, you can bet on the fact that I will probably no longer living in Arizona. I plan on forming a completely new family of people that care about me outside of my domineering mother. That is a relationship that is certainly going to burn out very quickly. Anytime she calls up and leaves a voice mail message telling me what she has been doing and wishes that I were back in Wisconsin, five seconds into the voice mail, I hit delete because she just goes on and on and on. She is completely in denial about the fact that things are changing outside of her control. What she is in complete misunderstanding about, is why I have no longer decided to focus my energy in the business world. I looked at this, and thought, this is not me. I already got a two year degree in accounting. Why am I doing this again? My life no longer extends itself to sitting behind a desk doing routine, boring, everyday type of work. With the things that are going on in my life now, and the things that are going on back home, there is little that me and my family have in common besides the same last name. I am probably the only person in my family that will also have a lot more money in the long run when things are said and done. No wonder they are scared for what is going to happen.

I used to be real close to my family. But in the aftermath of my hip fracture, and the reaction that I got from my "supportive" parents after everything was said and done has caused me to not only be upset with them for not being there on the same level with me, but also it has made it really hard to talk to them. At this point in my professional life, I'm not even regarding my parents as friends, because of some things that I found out about them and how they reacted to me by telling me to get a job. I don't see them trying to help me out in any way with the exception of stepping on my dreams, which I won't let happen in my lifetime. People that are my "true" friends help me out – by getting my work in front of people who can make a difference in my life. Such as those people that I met at my ten-year high school reunion.

INTERNATIONAL WINGS

Name one job, which involves travel as a regular part of the job. When things get to a high pitch, I will possibly be traveling to various places doing things that you could only dream about. If I get that promotion to creative consultant once my script sells to a production company, there is a good chance that I will get to travel around as the movie is filmed, possibly writing on location. At least I get first class transportation and somebody who will give me a meal and hotel allowance. That expense account is also in addition to my salary. I know there are those within my family who are going to say that I am not there yet, that's fine. Just don't come looking for me expecting some sympathy when you need something important, like money. Really, it is hard for my family to show any type of emotional support for what I am doing. If they think that telling me to get a job when our economy is already going to hell in a hand basket, they are deluding themselves. When my mother told me that her company was closing the doors because of bad financial health, I started to pray for a miracle to happen like immediately. It didn't happen. That is when I realized that I needed to force the hand, somehow. My phone was not even ringing regularly like it was before things got this bad. However, I was not picking up signs that things might be going for the better anytime now. They wouldn't get any better, because three months after my screenplay was requested by Keats Entertainment, it got passed on because of a lack of an adequate market. If the sales of this book have anything to say about the lack of a market, I think that I am going to get the last laugh on these people that wanted to say no to me.

My family would have no way to understand this. They can't understand why I get so excited about the day when I can pack the moving van and move to California. My immediate family does not understand why I need to have more people in my life than just them. My mother constantly wonders whether or not I appreciate the sacrifice that she is making for me. She knows that I appreciate it. Beating it out of me does not score points, however. It only pushes me further away. That is why I don't call her as much, and that is also why I don't answer her calls as much. What my mom does not understand is when millions of dollars are on the line, things tend to move a lot slower in terms of making an offer on my screenplay.

OUT OF THE CLOSET HOMEBODY

I spent a good amount of time in my hometown. It was the only thing that I knew. My family was really tight knit, pretty much staying in one place. Then, I decide that I want to get out and experience the world while I have a chance to do that. That is when I realize that I don't want to be part of my family's everyday life. It was twenty-five years too long. When I realized that other people that I knew from school that were branching out to different places around the globe, making new and different friends, and enjoying themselves, I had to have that same thing. Why spend life with the same people that you were born to? When I told my mother that I planned on moving to Phoenix, Arizona after graduating from school, she held a grudge against me and was very angry. (She will deny that as well. Go ahead and deny it. We both know who is right. I remember the pressure you put on me to stay where I was for the next millennium.)

The only person who I spent the most time with while I was still living in Wisconsin is my brother. He didn't try and control me. He would keep me company on the weekend and we would spend the weekend shooting the bull and watching different movies on cable. I would like to have him come out and experience what it is that I am experiencing on a grand scale. I think that when he sees that there is a life outside of the mother ship, that he will reconsider where he is in this life. Erik, brother, take this request very seriously. You need to get away from mom and dad. I know that I may be the black sheep in the family, but what I did was with the best of intentions. I think that they are beginning to become a bad influence on both of us. If you can't beat them, join them, as they say. A united front is absolutely necessary to prevent any more squabbles.

Sometimes relationships can be as confining as being stuck in a really bad marriage. Sometimes I hear people refer to their family as "toxic friends". Everybody has somebody in their family that is an absolute pain in the ass. I can name about two people who fit that description. After I moved two thousand miles to the other side of the country, I had changed so much that everybody in my family seemed like they were a space cadet. At least my brother accepted me for who I was. Don't laugh. It really is the truth – but only up to a certain extent. They still know that I may have changed on the personal level, but that is only on the surface. I am the same person underneath and easy to talk to.

There are people who don't want to let go, because they are afraid that change is going to make those that they care about do something that will enable them to go places that only could be dreamed about. When I realized I had something good coming my direction, I seriously considered directing as something to get into in the future. That way I am not only writing but I would also maintain the creative vision. That will come much later in my career, however.

That is what made me move to Phoenix after spending only a week and a half in the sunlight, which pushed the temperature well past the century mark. When I saw some of our neighbors who were doing just fine out here, I thought, maybe I could do the same thing. It has not been an easy road, but is life ever easy? Who have you been kidding? When I moved to Arizona, there were not only those that noticed the determination that it took to throw caution to the wind and move somewhere with absolutely no connections at all whatsoever. I must have been nuts when I did it. After I did that, there were people who commented to my mom that they wish that they had done something like that when they were younger – before they couldn't do that any longer. I possess that certain amount of spontaneity that is absolutely not to be tampered with. It is the essence of what makes me different than most people in my family, people who have, for the most part, decided to keep things routine. Before I moved, my mother was doing everything possible to delay me and divert my attention. I was on to that stupid game and was not going to fall for it.

That is what anybody with kids has to face. Someday, those kids are going to fly away from the nest, and home will never be the same again. They will be making mistakes, learning from them, and making it into something that helps them the next time. There is some serious truth to the fact that you can *never* go home again. As you change, so does everybody else relative to yourself.

THE AAR

The AAR was formed as a result of all of the abuses that literary agents use to mislead authors. If an agent is not a member of the Association of Author's Representatives (AAR), my advice is to stay away from them. Anybody worth the money – and their word – will be a member of this organization. There may be valid reasons for why they are not a member, such as if they are a new agency and don't have enough sales to qualify for membership, for example. A new agency perhaps has not sold enough books in the last year to qualify for membership. The general rule, the last time I checked, is that an agent has to sell at least ten or more books per year in order to qualify for membership in the Association of Author's Representatives.

I have a trusted source that I refer to if I can't find them listed in the AAR[1] database, which usually means that they are a new agency and have not sold enough books. On the Writer Beware's website, there is a source that you can use free of charge to verify if an agency is legitimate, meaning that they have actually sold a few books in the last year. You just type in the name of the agency or the name of the agent, and the database will tell you immediately whether or not there have been any sales by the agency that you are inquiring about.[2] By typing in their name, they will be able to tell you if you are dealing with a reputable agent or not. If the agency has had successful sales, I will be able to access that information.

The AAR has a lot of strict rules that members must abide by once they are accepted. This gives credibility to agents that they need because if they don't have it, writers like myself are going to tell them to get lost. Referred to as "the canon of ethics[3]" one of the things that is mentioned first and foremost, is that an agent cannot charge

---

[1] http://www.aar-online.org.
[2] http://www.agentverification.com.
[3] The Canon of Ethics is a rule of behavior that says what agents are forbidden to do, such as charging upfront before a sale is made on a proposal or manuscript.

money for anything upfront before a successful sale is made to a publisher. The only thing, as a general rule, that an agent can bill you for are pre-approved expenses that are agreed upon between the author and his agent.

That is why I am strongly in favor of the AAR. An agent can hold an author responsible for a reasonable amount of expenses for the purpose of photocopying, postage, federal express mailings, and other minor incidental expenses, but they are **_never_** to collect that upfront and claim to deduct that upon making a sale. If any agent uses that kind of language, run. Those agents are stupid idiots. These agents are crooks. Sometimes, however, after trying, even the most competent agent cannot sell a proposal to a manuscript. My friend's agent sends authors a bill if she is unable to sell a proposal within a year and most authors gladly pay knowing how hard she works to sell a manuscript. But for me, I don't know if my material was even seen.

The moral of the story, is that before you send anything to agents, you have to qualify them, and make sure that they are legitimate. If I were never woken up in a rude way in January of 2004, I never would have learned about this. The only truly legitimate agents are members of the Association of Author's Representatives. That was my mistake the first time that I started in this business. I was not qualifying the people who would be doing work on my behalf. There are so many scoundrels and idiots out there, that I could write an entire book about them. They are not limited to books, either. As somebody who had already been through this once, I knew what was going on when the agent who I had contacted asking to read my script, who I won't name, asked me to pay him $500 to do a "critique" of my screenplay was doing nothing except dicking with me. He tells me to send the first twenty pages and he will "evaluate" them free of charge. What he was planning to do was reel me in, hook, line, and sinker as though I was a complete idiot. Do I ever make the same mistake twice?

Some people will ignore this advice and take any agent that comes their way, just for the sake of saying that they have an agent. People who have been struggling, want to make it look like they have everything in order. This is a bad thing to do. Having a "business card" agent is worse than not having an agent at all. Being able to say that they have an agent makes them feel like they are getting somewhere.

A real agent should send you copies of rejection letters that they receive. Being new, I did not understand the importance of this concept. I was not getting any correspondence with regard to rejection letters, which should have been sent to me. My agent would not even send them to me – perhaps because she was not doing anything with my material. Every time I would ask why she doesn't give me a copy of the rejection letters, she replied that most of the letters don't really tell an author why the book was not accepted. Maybe she was telling me this because she really wasn't doing anything on my behalf. The whole concept here is open communication. With the agent that was going to try and do a critique of my screenplay, I requested to see the language in his contract. He said something like, he only uses a contract when necessary. If an agent tries to work with you and they don't sign a contract, run. I am not an attorney, but if there is no contract, there is no commitment by the parties to be bound by the contract. At that point, it is a verbal contract, and very difficult to enforce in court.

Everything comes down to communication. If the lines of communication are not open, and your agent is not keeping you informed about what he or she is doing on your behalf, start looking elsewhere. Otherwise, if I don't do a good enough job, the project is shelved unless another studio buys it off in the situation where the project goes into turnaround[1]. However, the general goal is for studios to purchase viable projects that won't go into turnaround. A lot of successful movies such as Steven Spielberg's *E.T. The Extra Terrestrial,* was one of the movies that was bought in turnaround and set up at another studio.

Another little thing that my friend would share with me was how long it took her to find an agent. She said, that it took her five years to get an agent. What I was never advised to do was submit articles to magazines to get noticed. That also gave me something to do while I figured out how my movie was going to pan out. My friend would point out that if an agent is interested in your work, they won't write a nice letter saying how good your work is. They generally call you. Well, I tried all of the above and the phone isn't ringing.

Anything you get through the mail is negative – for the most part. Sometimes there are reasons for why an agency can't take your work, such as changes outside of their control. Most of the letters, however, would say that their client lists are full and they aren't signing any new writers. Almost all agents are servicing anywhere from

---

[1] When a project doesn't make it out of development, outside producers can purchase the project by paying everything that is against the script and get the project set up at another studio.

twenty to thirty clients at any given time. In reality, that is a lie. Any agent with a brain knows that if there is a writer with a hot spec that just made the trades with a huge sale, that maybe that person is worth signing. After all, would any agent worth his or her mind give up ten percent of $500,000? I guarantee you that they would be doing anything to get that writer's business. If I thought the phone was quiet now, it will be the exact opposite when word of mouth gets around about my screenplay selling and my name is in the trades. Then agents will be trying to make me sign on the line.

If you are in doubt about an agent, ask a producer what their experience has been with agents. A producer won't lie to you about that. At the time when I was trying to get read, I asked another producer who was reading my screenplay, but passed on it, whether or not it was ethical for an agent to charge money upfront before a sale was made on a screenplay. I was asking this in regard to the "agent" who told me that my screenplay needed a critique. He said that no agent has a right to ask for any money prior to a sale. Producers won't lie to you. They will be absolutely truthful about an agent's track record, if they know it. The reason for this, is because agents cannot be producers because that is a conflict of interest. You can't buy low and sell high. So, there is no reason for a producer to be withholding information about agents. They are getting material sent to them by agents all the time and know if an agent is reputable or not and after a while start recognizing the names of agents. Plus, this agent, who said that they couldn't attach their name to my project because of the way it was structured, in reality could not do it because producers who are signatories cannot read any material submitted by agents who are not signatory to the guild. Take my word on this.

In my search on the internet for literary agents, I came across another agency in Texas that I had not seen on the internet before. Hmm.... This caught my attention, especially with the San Angelo address. The website contained the same advice that was given to me by my ex-literary agent, who has now vanished into thin air. As of the writing of this book, detectives are still looking for her, and as far as I know she has not yet been prosecuted. By the time this book is on the shelves, she probably still will not have been caught.

Supposedly, this agency was opened by an employee who worked for my agent. It made me reconsider everything that was told to me by this scoundrel. It seemed like everything was being told as a lie to me, as a way to cover things up. I had no idea what she was doing, but one thing was certain. I was not getting any responses from publishers. As a matter of fact, the publishers who were aware of this, sent her a request not to send any more material.

The fact of the matter, is that you have to background check anybody you are sending your work to these days. People get burned every day by these crooks. I got burned. And it wasn't the fact that I got burned that was frustrating, but it was the fact that I had a year of my life wasted by this person claiming to be a literary agent. Informed that my manuscript was in police custody, that was when I was pinched awake and brought to reality. Somebody burst my bubble, and I was not very happy about that.

Even worse, I was fed lie after lie after lie by this agent. She strung me along for an entire year and spun nothing except a web of lies and deceit. She said that my work had been approved by everybody in attendance. She even lied about the fact that she liked my work, saying that she was happy to receive my work, and slyly admitted to the fact that there are things that are in this contract that were not previously talked about. This is what I am talking about when I say "open communication". To add insult to injury, reports that I found on Writer Beware was that all the titles that were on this agent's website were published by vanity houses and were not really sold by her office.

The exact wording in the letter, which I still hold in my files for legal purposes and other reasons, was the following sentence: ***"I realize there are aspects of the contract we have not discussed previously. If you have any questions at all about the enclosed documents, please do no hesitate to call."*** Well, if I saw something like this after being burned like I was, I would not hesitate to throw it in the trash. Because that is what that document is. Nothing but trash. Here is the exact text of the letter that I received from this agency:

Dear Author:

Congratulations on the quality of your material; however, the evaluation letter you have received is not an offer to represent...it is an offer to discuss the possibility of representing your work.

Each month [we] receive more manuscripts than we are able to handle. Most receive polite rejection notices...either for the lack of quality or an inadequate market.

Very few receive a positive review such as you have...and from those, we must make a decision on which ones will actually receive a representation agreement.

To determine exactly which manuscripts we will offer to represent, we hold a committee meeting on Thursday of each week. The committee has as many as six members (depending on who is available at the time.) Each person on the committee reviews segments of your material, looking for quality and skills. We look at not only how many publishers are buying your type of material but specifically which publishers...what kind of history the agency has with those publishers...what kind of current relationship we have with those publishers...etc. We also look at the last time we took on a project like yours...did we make money...did we lose money? The primary purpose of the meeting is to draw a conclusion about the probabilities of being able to market the material and earn our 10% commission (our bottom line).

It takes a majority of the committee members in attendance to vote in favor of your material in order for us to offer a contract: i.e. if all six members are present, at least four must vote in favor of your work. If this happens, a contract 'offering to represent' will be prepared, and sent to you, the same day...along with all accompanying documentation.

If less than a majority vote in favor of your work, we will not offer a contract; but, if you have provided a return mailer with sufficient postage to handle it, your materials will be returned along with a brief explanation as to why it was not accepted.

The decisive factors for offering to represent are: The material must be of sufficient quality. If criterion is met, we will offer...if it is not...we will not. To make it marketable, there must be a viable market. We must believe the person on the other end is someone we can work with for a full year without conflict.

When I read this, I got false hopes. Part of this was my own denial, because I wanted to believe that I had a literary agent who was there for me. As it would turn out, having a "business card" agent, is worse than having no agent at all. Because this person is misleading you, toying with your emotions, and knows that you are vulnerable. What I needed was a publisher, instead of an agent. The agent would come much later. These agents have no selling ability whatsoever, and have to charge upfront marketing fees just so they can put food on the table. Which is what I thought that was all about. I thought, if

I were an agent, how would I bridge the gap so I could at least avoid starvation? This thinking that I had clouded the issue. Unless you are dealing with a writer that closed a huge deal, it is hard to live on ten percent. When I saw her listed in Writer Beware, which I also was in denial about, that clouded my vision even further.

First, anybody can post anything about anybody on the internet. I didn't know how reliable this information was. It could have been posted by somebody in retaliation for being turned down. Then the day of reckoning came. In January of 2004, a search warrant was served on her home and business (what business?!) and my manuscript was put into evidence, which immediately lowered the value of that property.

No agent was going to even touch a manuscript that was in the possession of the sheriff's department. The real kick in the butt, is that even if a reparation order is made, that almost all the time, there is little chance I will ever see the money that I sent her, even though I made a copy of the front and back of the check that I sent to her. I thought, well, if I can't get her for stealing, I'll nail her for tax evasion. Because chances are, she wasn't even reporting that in her income when her tax return was filed. With changes afoot, however, I had better things to do than chase after an amount of money that was the equivalent of working for peanuts. And I also have problems that are going to be a lot worse than being burned by this agency.

The reality that I was in so much of a hurry to get my story out to the world, that I blindly submitted to this agent on the internet, and didn't have a reason to think that they were dishonest. Only then, did my good friend tell me that it is unethical to pay any money upfront before a sale is made. Why is it that my attorney could get away with doing that? I had to pay my attorney a $500 retainer fee upfront and they are very similar to agents in terms of the work they do for their clients. And, there is no guarantee that my screenplay will even sell. Both agents and attorneys have the power to negotiate deals on behalf clients. A lot of agents are also entertainment attorneys, and that is where the line gets blurred. Why is it that attorneys can require a retainer fee, in the form of an upfront fee, but an agent can't? Despite this life experience, which I will never forget, that is still mind boggling to me.

The best way to avoid this type of situation, is to only deal with agents that are located in California or New York. Since agents are located where the action is, that is better anyways. It doesn't do any good to send your script to an agent in Texas. After my experience, if I see that an agent is located in Texas, especially San Angelo, that is a red flag to me. Plus, agents in California are subject to licensing requirements that other agents are not subject to. I should have been suspicious, just because the agent was located in central Texas instead of where all the action is located.

These "business card" agents who charge money upfront, have to charge fees because they lack the ability to be a real agent. All of the titles on her website, which she claimed were agented by her, were supposedly published by vanity houses. She robbed me and countless others of our time, emotions, and hard work. I was trying to get a message out to the world about the link between epilepsy and osteoporosis, and my life experience, and she was doing a good job screwing things up for me.

The aftermath of the story here, is that I did not know how I was ever going to get another agent, especially after my material was sent to numerous publishers already. A publisher is not going to give something a second look. Getting another agent to represent me would be as easy as getting another job to replace the job that you were fired from. And the agents that would consider me, would question my authority to talk about a topic that doctors were unable to address. Thinking that being through this on a first-hand basis, that would give me some leverage, I was wrong. The thinking was, if doctors are unable to understand it, how could I understand it?

Once a publisher passes on something, you can't submit it for a second look. That is why I have to keep records of who received my manuscript in my database. If I get another agent, I will have to disclose this information to them. They are not going to like looking like a ditz because I did not keep them informed and keep the channel of communications open. Something I have found through my life lessons, is that there are certain things I have to disclose, even if it means that I am putting my livelihood at severe risk. If I don't disclose that my manuscript has been possibly seen by numerous publishers, even though I don't lie, it is a lie by omission. At the time, when I was submitting to these publishers, what I didn't know was that she was the one who was supposed to do all the administrative work for me. That made me question at a certain point, why it is that I am paying her ten percent of my earnings if I have to schlep all the manuscripts to the post office. Chances are, that anybody that saw something come in from her address did not even read it.

EUPHORIA

The fact that my first manuscript documenting my life story got trashed in value due to being taken into possession by the sheriff's department was the shocker that bumped me out of the comfort zone and into the area that touches celebrities. That is what got a movie written about my experience.

Indirectly, the fact that I had a dishonest agent to begin with made my name known by more people than I had realized. I had to find a way to get people to come into the bookstore and ask for my book. What had started as a web page for my book became a full-blown franchise without me having any control over the monster. It consumed me, according to words from my mother. From what I read, if one of your books bombs, good luck finding another publisher to publish another book that will be a waste of money. You are, in effect blacklisted unless you write a book under a pseudonym. Just like everyone else, publishers are concerned about their bottom line. With the advent of computers, all a bookstore has to do to determine the sales volume of your last book is type the name of your book into a computer database and they can see instantly how many books have sold in the different bookstores. That is how bookstores make their decisions on what books to order. If bookstores won't order your book, why would an agent want to represent you?

After I realized this, I knew that drastic measures had to be taken to keep my name in the spotlight. Short of hiring a publicist, which I will probably still have to do, I wrote a movie about my experience, which ended up generating a lot more interest than I was used to getting. But, I could not focus on that, until I had a better idea of what I could ideally expect as an outcome. The life story rights and movie script alone could easily go for close to a million dollars, even before the salary that I might earn as a creative consultant are thrown in the equation. My attorney is working real hard on that one when the time comes, although at this point, there are no guarantees.

I don't know what these potential publisher even said about my material, because of the fact that all of the correspondence was going directly to my ex-agent's office, which is now abandoned. That is the worst part. Based on information that I obtained after the investigation ensued, there is a good chance that a lot of time, the work was never even read, but again I have no way of knowing what really happened – because there was a lack of communication from the start. Almost all the time that I would call, her line would be busy – and my phone calls would not be getting returned.

There had never been any open communication from the start. The web of lies told by dishonest agents can be enough to put a person's professional reputation right on the line, which is the equivalent of a professional death card, almost like being blacklisted. At this point, I was literally sick with worry about everything under the sun. With the numbers of visits to my website taking a severe nosedive in the aftermath of a search warrant being served on my agent's home and business, I was only thinking about my own well being at the time. To make matters worse, I thought about what else could happen. I was fearing that I would get a court summons and have to testify in court. Knowing the way in which the police operate, what would have happened if instead of pursuing her, they decided to come after me, since I was associated with them for over a year? I had just taken a class in accounting and my professor hands out an article that talks about how following orders by your boss is not an excuse – in effect employees are told to perform make journal entries to a company's books to make the bottom line look better. In the process of doing this, they end up doing time in jail for a very long time.

In March 2004 I was on the phone to my father when he was in Anaheim, California. I was choking. I feared spending a good amount of time in the gray bar hotel. I remember talking to my dad on the phone was overly concerned that I was going to get summoned. Worse than that, I did not have any clue about how I was going to revive my freelance business. I had a movie script, which luckily was never in her office, so I spent a good amount of time on marketing that – only to be told that no market exists. Well, let's see about that. When they see the number of people that request my book, perhaps they will change your attitude.

If Leonardo DiCaprio is reading this book, I would really like you to star in the lead role. Based on your last movie, *Catch Me If You Can,* I walked out of the movie theater inspired for the first time in my life.

Jail is not where I want to be. I want to be where I can tell my story. And that is where my battle to get my story told really starts to unfold. With no agent representing me, I was starting from square one.

HOLLYWOOD

According to some sources that I read on the internet, at the stage that I was at in my career, I didn't need an agent. That would all come later, like when I sell my screenplay to Hollywood for a lot of money. Agents don't want to set that fire; they only want to set a wildfire off of the heat generated by the writer by saying that you are the best thing out there since sliced bread. I was led to believe that I had to have an agent before I could even sell my book. I could have easily gotten a publisher first. I just didn't want to wait fifty years.

I had been counting down the days until I would be able to drive to Beverly Hills and meet with a producer and talk about how to improve my screenplay beyond its current stage. With the number of times that I had been submitting my screenplay to a script coverage service and being told to change something else every time the next round of notes came back, I was starting to notice a pattern that I did not like and decided to take the bull by the horn. If somebody was going to tell me what should be changed, it was going to be in person, or they were going to hand me a check and pay me to make the changes. I was no longer planning on doing any more free re-writes to satisfy industry professionals.

I was getting to the point where I was re-writing myself to death. I had things just perfect – and if I was going to change one more thing, I wanted to be paid first. It gets to a point where you can only rewrite so much. I made a note to write a letter of complaint to the script coverage company, basically lambasting them and blaming them for all the dumbed down movies that exist presently. In their report, they even said that I should overhaul the script so that it would be something that everybody could care about besides those who had a disability. At this point, I put my foot down. Hard. I knew based on surveillance reports that I ran off my website, that they were lying – I knew that there was plenty of interest in my story. There was no way that I could convince them of this fact, however. The only way to convince them of that fact was to get this book into print very quickly.

Everybody that I talked to in regard to my story was curious. How is it that a twenty-six year old male finds himself diagnosed with severe osteoporosis – and finds his life changed instantly? The movie script itself is only one aspect of what an entertainment company buys. What they are really buying, is the pitch and the logline[1] for your movie, which is one or two sentences which summarize the entire movie. What a logline basically does, is tell the studio what the movie is about. Believe it or not, the logline will tell the studio whether or not something is going to be too expensive to produce.

Some of the things the script coverage company said made sense. But they wanted me to rewrite the entire screenplay and involve a completely different premise. I'll do that when I am hired to write the sequel, thank you very much. After what I had been through with a dishonest literary agent, the script coverage company was already on thin ice with me and I was not somebody who would have a problem letting them know that they were also full of shit. I can't wait until the proof is in the pudding – because then I will be writing that nasty letter to them – and enjoying every damn minute doing it after the unnecessary torment that they put me through. Really, there are some people in Hollywood that enjoy torturing writers. This company is one of them.

They basically told me that either I rewrite the entire screenplay or I put what is in the screenplay in a different medium. They said that various parts of my screenplay should be deleted. When my script sells, I will post those notes on my website. The reason I did not put them up there originally, is because when I read the notes, I was offended and I did not want that to influence the sale of my screenplay. These notes also contradicted the notes that I had gotten previously. All the work that I was putting into this screenplay was not making them happy. It was perfectly good material. I knew it. They knew it. They just wanted something that would be picked up by every producer under the sun and get writing jobs for people in Hollywood. It's time for me to blast this script to every producer under the sun and see what the script coverage company is really trying to do to me.

THE FULL BLAST

After getting their report, I was frustrated with the whole damn process. I, again, throw caution to the winds and try a different approach. That is when I used Scriptblaster to see what type of results I would get. Only two hours after my query went out, I got an

---

[1] One to two sentences which summarize the theme of the movie. Most movies are bought based on the logline alone.

email back asking to read my script. That says a lot for the "coverage[1]" that was done on my script. After seeing my business almost nosedive to the ground, in the aftermath of being notified that my original manuscript was in the hands of the sheriff's department in January of 2004, that is when I shelved the book project and focused entirely on writing the movie – and becoming sleep deprived. My goal was to get everything back to where it was in February. My original manuscript had lost value because of who had possession of it.

Based on website reports that I ran in July of 2004, the number of hits to my website went back up to the same level that they were at in February. This came from a lot of sacrifice. There would be times that I spent about eighteen straight hours doing nothing but working on my computer.

I was not somebody who had a problem when it came to running my mouth about something that I though was not in the best interests of the person paying for the movie ticket. In my opinion, I was writing my script based on what the average person would like to see, not what a producer would like to see. There are various reasons for why a studio will decide not to request a script. If you ever write a movie, you will find out how much risk is involved with actually making the decision to purchase a script – and then watch it bomb at the box office. Again, this has to do with one thing. Protecting the status that a studio executive has in the industry, and then explaining to the stockholders why they bought a movie that ended up bombing at the box office. People end up on the street because of these decisions, which is why they are decisions that are not made lightly.

AN ALLY WITHIN

Beverly Hills, California – March 19, 2004 – I had been submitting my query letter to probably a few hundred production companies with no takers. I would later find out why. I was also getting dicked with by the script coverage readers that I was sending my new and updated script to. Every time I would fix what they said was wrong, they would find something else that was wrong. Every time I resubmitted my work, I sat and waited to see what other bullshit they would feed to me as if I would be dumb enough to go along with the plan they had in store for me. At a certain point, I had decided that

---

[1] Coverage is a report done by readers at a studio that tell what a movie is about. Coverage also recommends whether or not the movie should be exploited.

nothing was going to get me a "recommend[1]" and get my script closer to production. Although some of the notes gave me some pointers, the way I saw this was that somebody had to justify their job by passing on already valuable material. Whoever said that everyone reading my movie script was right in the first place? Scripts that are evaluated by two different companies can get wildly different coverage because of who is reading the screenplay. From my experience, most of the time coverage done by production companies is along the lines of what movies they make. Agency coverage, from what I have read, is more forgiving. Agency coverage is more along the lines of whether or not the writer has a clue.

Reading on a website that a producer was agreeing to meet writers one on one and go over their script, I saw this as an opportunity to get my work up to the next level and possibly sell. I tell my parents, who, don't forget, are supporting me financially, maybe not emotionally, and my mother goes to work drawing up catastrophic things that could happen – I'll get hit by a train without even knowing it, I'll get my car wrecked, things that stand a chance of happening on any day that passes by. Hell, with my insurance rates, I don't need another accident. I resisted these scare tactics with all the power that I had and slammed the phone down. Hard.

I went on guts instinct and decided that the reason the script coverage people were passing on my script all the time, was to get my material so diluted that any producer in Hollywood would want to exploit it in their own little way. I was not going to sacrifice good material. I came to the conclusion that a story like mine would never remain in its original form if submitted to a script coverage company. The goal of these companies is to find marketable material that can be submitted to a copious number of producers. A majority of the producers don't want to look at material that has life story rights[2] attached to it, which thereby, raised the value of the property. That was part of my problem. I had to remain confident that my script was the best that it could be under the circumstances. I believed in my script. It just had to find the right home where somebody would want to exploit it.

---

[1] A script that gets a "recommend" is a movie that gets passed to producers. However, not all movies that are bought will ever be made into movies. Only a handful of movies that are bought end up actually getting made.

[2] Life story rights are separate from the actual script sale. A studio can buy life story rights without buying the script. However, if they buy the script, the life story rights are extra.

My individuality was getting in my way and I had to shut my parents out of my life. I knew that I was being watched over and nothing would happen to me. All this amounted to were blank threats to keep me from finding out more about what I needed to do to improve my situation. Never mind the fact that this is coming from the same person who is helping me out financially. I had to mentally prepare myself for this trip that I was about to take. I was a pro at survival – I had taken a two thousand mile journey three years ago before I actually moved to Arizona, so I was not fearful at all of something happening.

Interstate 10 – March 19, 2004 – At 4:30 in the morning, I left my sanctuary in Phoenix, Arizona and dashed off to California. The sun would not be up for another two and a half hours, and I wanted to beat the rush hour traffic, getting a head start while most people were still sleeping. Setting the cruise control, I begin my journey on I-10 about 375 miles all the way west with two full cans of Pepsi in case worse came to worse where I out of nowhere start to doze off. I was getting excited. For the first time that I have been living in Arizona, I was making a trip to Los Angeles – the place where everything happens in this country. Stopping at a McDonalds, thirty miles out of town, I wished I would have done this before I got on the freeway, as the prices were off the chart. At around 6:30, it is still dark outside with no sign of civilization in sight, and no radio stations are tuning in.

Glancing at my cellular phone, which is sensitive to changes in location, much like a GPS tracking system, it is able to tell me whether I am in Pacific Time, in which case my parents are three hours ahead, or Mountain Time, which means they are only two hours ahead. The good thing about my cellular phone, is that I have a nationwide plan, where anywhere I go, as long as I am in network, I don't have to worry about roaming fees and am able to search the internet using my phone. To save about $100 a month, I dropped my landline phone. Why pay long distance charges when I get the same service for free on my cellular phone? This decision would turn out to save my sanity a year or so down the road – around the time that my story was a hot issue around Hollywood and California in general.

Something interesting about Arizona, is that with the exception of the Navajo Indian Reservation, which is in the northern part of Arizona, it was decided that setting the clocks ahead one hour in the spring and one hour back in the fall was not worth the time, especially with the three hundred days of sun that Arizona sees every year. Instead, we just follow what the rest of the country does. In Arizona we already get more light than most regions of the country, so there is no need to recognize daylight savings time. Out of boredom, I give my father a call and tell him how things are going. I find out that I must be going through the Mohave Desert. At 3:00 in the afternoon, I arrived in California and settle into my hotel. and waited until the following morning for the complimentary breakfast. I get to my room, zapped from six hours on the road and sleep until about midnight, when I wake up and, out of boredom, surf the internet on my laptop computer, using my cellular phone which has connection speeds that are about as slow as the traffic in a California rush hour.

I email my parents and tell them that I made it safe and sound. When my mom was out here with my father, she was probably getting bored and found things to worry about. (You know, mothers instinctively find things they need to get their children worried about.) She made it seem like the traffic was worse than it really was. The traffic was not much worse than the rush hours in Arizona and I had to deal with those regularly when I went to school after I finished up with my job every day. Sometimes the traffic would just creep along moving at about the pace of a snail. When that didn't work to sway me, she tried to get me to think that somebody was going to kidnap me for financial gain, at which point, I slammed the phone down. There was nothing I could say to her to put her at ease. Ever since I originally moved from Wisconsin, I would no longer be the same person that she had known for so long. There is an element of truth to the saying, "You can never go home again." Well, you can go home, but you will be completely different from the way in which people used to know you. A lot of things change over a two-year time span. She wanted me to stay in my apartment, so that I would not know what else was out there for me. I decided what was important to me, and what was important to me was cracking through and making a financial success, not pleasing her by any means. I was in no mood to compromise with a hostage negotiator.

With prices off the chart, I realized that everything in this area was extremely expensive. Somebody was going to have to pay me a lot of money to force me to think about moving here. It seemed like the prices were past the point of inflation. Just a one night stay at a hotel cost me well over a hundred dollars, which I could barely afford being on a tight budget, since I had to incur other important, but necessary expenses while I was here in California.

INTERSECTING PATHS

About a week before I planned on making my trip to Beverly Hills, my father was out in Anaheim, California a week earlier speaking on behalf of a report that he was writing for work he was doing in his field as a social worker. It was interesting that our professional lives were both taking interesting turns around roughly the same time. It would have been interesting if we had both been in California at the same time. At least I would have had something to do outside of work-related stuff.

I was traveling to Beverly Hills to go over my movie script with a working producer in Hollywood, because my efforts were going unnoticed by a lot of people. At 1:00 the following afternoon, I walk into the coffee shop where I was going to be meeting him and am immediately recognized by somebody who must have seen my website. I don't know for sure, but I didn't know them, although I was certainly playing the part with a suit and tie on, especially since I was meeting with my attorney later on in the day. The idea was to put on a good impression when I met with him, as I knew that my attorney was going to be a part of my long term agenda when my screenplay got close to a sale.

THE NEW KID ON THE BLOCK

I couldn't get read by agents. Agents have a problem, for some reason, taking on new writers. The ones that did want to represent me, wanted to charge me a premium upfront, which I was not going to pay. I was hoping that this meeting would get my script in the door somewhere. As it turns out, it would have been turned into scrap material, since it lacked the dramatic shape that it needed to have to make it a successful drama. I was experiencing frustration with all this. What was it going to take for me to finally break through and hit the big leagues?

The producer made suggestions to me that I be really descriptive in the way I write individual scenes and even provided examples of how to do this. For the money that I paid him, he spent roughly two hours with me and got through most of the script. I had previously been submitting this script to a script coverage service and they kept passing on it for one reason or another. That is what motivated me to pay top dollar to see this producer. For all the information he gave me, I was paying him roughly $125 an hour for information that I would not get from the script coverage company. Some would argue with me, but this was money well spent. I am glad that I took this opportunity while I had it. Otherwise, I would be stuck doing free rewrites for the script coverage company, which I was starting to mistrust.

THE FINANCIAL FOUNDATION

After the producer wrapped things up, my next stop was in Santa Monica, just ten short miles from where I was at the time. I was meeting Joey Sayson[1], an entertainment attorney, to sign a retainer agreement and get questions answered with regard to the process involved with getting a motion picture developed. The only problem was that absent a deal being on the table, he couldn't specifically tell me what it was worth. He could only give me a ballpark figure of what rights he could exploit in my favor.

Santa Monica, California – March 19, 2004 – On the same day, I met my attorney, Joey Sayson, the one person who was going to save my life from a chaotic train wreck. He was a person who would serve as a substitute agent, since I could not get read anywhere. With his permission, he allowed me to list his name as an agent on Inktip.com[2] since he was performing roughly the same role as an agent would be performing. The main difference, was that it was my responsibility to find an interested party and get an offer for my screenplay.

Once I found a home for my script, my problems were going to be much easier – and again – also more complicated at the same time. For the first time, I had to consider the fact that I would have to pay a personal manager to get my business going so that I could run it effectively. If things were difficult now getting work, they were going to get more complicated when I would be going on a nationwide tour promoting my work to the public to increase public awareness of what I was doing – as if they didn't already know. If the numbers of people who I didn't know that were approaching me and addressing me by first name said anything, it would be that I was becoming well-known without even

---

[1] Joey Sayson, my entertainment attorney, works in Santa Monica at The Law Offices of Sayson and Associates. http://www.saysonlaw.com.

[2] http://www.inktip.com. This website is for screenwriters and allows them to put their screenplay, synopsis, and treatments on the site for producers and other Hollywood agents and executives to view.

knowing about it. This would be a scary situation for anybody. In Chapter six, I talk about some of my experiences with what is known as the publicity factor, and it is not something to make light of. It is something that makes anybody who is well-known concerned for their own security. Don't take it personally.

Although my attorney was not doing the same things as an agent normally would do, he was in effect playing the same role by negotiating the deal. Since I was in town, I had set up a meeting because I like to know that somebody is more than a voice on the other end of a phone. He went over the rights he would be able to exploit for me for my screenplay, which included the actual script sale, which as outlined by The Writer's Guild of America, specified that the price to be paid for original screenplays, at a minimum depending on the budget of the movie, were to be no less than $50,000 for a low budget movie, or $100,000 for a high budget movie. One of my questions, obviously was how much could he get me in a perfect of perfect worlds? In addition to my spec screenplay sale, he said that he would try and arrange a payment of life story rights which would bump the value of my screenplay up even higher. My guess, with the original story that I have, is that I would be paid close to a million dollars or more, depending on the mood of the production company on a given day.

He even said that there existed a possibility of arranging a salary for me as a creative consultant where I would meet with the director and the individual writers to share my experiences and the direction I saw the screenplay going. At least that will give me a salary in addition to my contract.   Once I sign over the copyright to my script, the person who owns the copyright has the creative privilege of fictionalizing it as much as they see as necessary for the movie. With executive producer compensation[1] and all of the other perks, my attorney envisioned me getting roughly $400,000 to $600,000 in total. Certain elements could bump this figure higher.

Essentially, I was creating what is often referred to as a "work for hire[2]" where what I create is actually owned by the studio, since it is created under their auspices. Although I create the work on my own time, I may actually employed by the studio

---

[1] The **executive producer** is a person who is responsible for bringing a project into the studio. This is also the same person that claims the award on award night for best picture.

[2] A **work for hire** is a work specifically commissioned by an employer. Even though the work is created by the employee, the author of the end product is the employer.

through my loan-out corporation[1], Trimberger Enterprises. Unlike with this book, where the publisher has to get permission before even one word of text is changed, with my movie script, they can keep or change anything that they want for the purpose of making a movie because they own the copyright. As a matter of fact, when you see the movie in the theater, chances are it had already changed in form from the original material, since a number of writers work on the picture.

The original spec script is, in a way, the foundation for a movie, providing something to work with to create a better movie. There are times that something in a movie script is not actually used in the movie even though it was in the script originally. In effect, what is contained in a movie script is never the sole work of the original screenwriter – although the original writer is given credit despite the fact that others may have re-written the work according to what the studio wanted at the time.

One of my biggest concerns was the minimization of a lawsuit, since I was dealing with characters who were real life characters. In the meeting with my attorney, he said that one of the things that he would be sure to do is to name me as an additional insured on the errors and omissions[2] insurance policy. *(Note to self: Make anybody who you think stands a chance to sue you sign a general liability release agreement.)* Actually, before an errors and omissions policy is even issued, it is my responsibility as a creative consultant to secure the releases of anybody who is depicted in my movie. According to numerous sources that I have read, every movie that has made has gotten sued for one reason or another. A lot of the lawsuits in Hollywood pertain to idea theft – someone just makes a seven figure deal on a movie, and then a lawsuit appears out of nowhere saying that the idea that I saw originate from day one was someone else's idea. In case you are writing a screenplay, if you think that you have a clever twist on an idea that has never been made, I can guarantee you it is either in development somewhere or somebody else is writing it. Ideas are about as free as the air. What matters is the *execution*[3] of the idea.

---

[1] **Loan-out corporations** are for people that make a certain amount of money per year. It operates to give a tax savings to those who are employed directly by a studio. Loan-outs function to "loan out" the writer's services to the studio, serving as an intermediary.

[2] **Errors and omission insurance** is insurance which protects the writer and producer from claims of copyright infringement, libel, and any other lawsuits associated with the project.

[3] The way an idea is represented in a script. Ideas themselves are not copyrightable because they are not fixed in a tangible medium.

When I met with my attorney, my only questions pertained really to how much things would be worth, what type of work he could arrange for me, and if there was anything he could do to help me. He went over what he could or could not do, and who he would refer me to in the case of litigation, which he does not handle. I am sure that I will have a lot more questions when I sign the contract, which will almost be as large as my movie script.

Interstate 10 – March 19, 2004 - On my way home to Phoenix, Arizona, I started dreaming about breaking it big, and using my original story to jump expectations and actually get paid a seven figure sale for my movie script. Although in actuality, this is rare for a first time screenwriter, it didn't hurt to think that this million dollar lottery could actually happen. After all the activity, I got dangerously close to falling asleep behind the wheel doing 80 miles per hour. Wherever I was, there was no sign of civilization. It got to a point where I had to do circles in the gas station at slower speed so I could get out of the time warp. The high speeds on the freeway for long periods of time were enough to make me go crazy. The high speed on the freeway for a long period of time was making me nauseated and dizzy – and the last thing I need to do is crash my car at a high speed. My head is doing circles and my eyes are struggling to stay awake. At this time, I am still a good 100 plus miles outside of Phoenix and would not get home for another two hours. As I got within ten miles of home, I am barely able to see straight. I got back home at 5:00 the next morning, and just used the entire day to unwind. I was refreshed knowing that I had some new knowledge to apply. What I really needed was a sale for my screenplay.

LIVING LIFE ON THE EDGE

Due to the fact that I was writing a screenplay, I had educated myself about how the system worked and what to expect in terms of what my creative rights as a writer happen to be. Every time that I would talk about this to my friends, I would get no response. I was talking about something that they had no clue about. I had isolated myself by becoming different. Because I was not the same, and not doing things the way that most of society would do them, I was looked at with suspicion. At one point, my mother, who is currently one of the only people who supports what I am going through, asked me what I am doing and actually visiting my website from time to time to show an interest in the progress that I have been making. One time, when I was visiting, she gave me this postcard, which I still have somewhere in the mess of things in my apartment, which says, "If you are not living on the edge, you are taking up too much space." I got a good laugh out of that. If you think in a statistical sense, ninety-five percent, or the mean of a population, is located in the center taking up a majority of the space. Five percent or less of the population live their life on the edge, taking risks that most people would not take.

The people who take risks are not given the full credit they deserve. They are misunderstood, because they are not like everybody else in the population. How do you think Bill Gates and Steve Jobs started out? They had money, sure, but they couldn't have gotten where they are today without support from somewhere. Steve Jobs started making and selling computers out of his dorm room in college. Later on, he started to take that to a grander scale and started what is known today as Apple Computers – the type of computer that my father has sitting on his desk. He is one of those individuals who lives life on the edge, having a computer which is not compatible with what the majority of the population in society uses. I think he does that because of his dislike of Bill Gates. On the other hand, while I followed his line of thinking, I switched to IBM, because that was what everybody was using – and in this world, you sometimes can't afford to be different – especially since all the schools have IBM instead of Apple computers.

That is when I came upon another realization. That my friends, although supportive of me, had no way of being able to understand the language that I was speaking to them. Every time I would talk to my friends about my movie, it was like I was speaking another language. Because that was dominating my life at the time, I felt that I had no choice except to cut off all communication with them, because it was difficult enough getting them to understand things on my level. It began to hit me close to home that the people who I had formed friendships with, I was no longer able to be friends with because my elevated status made me a completely different person – outside of my own control. There are prices associated with change. People who you had known a year ago change into completely different people. The way I looked at it was, if they are unable to accept me for the person I have changed into, it's their loss. I am not going to waste my energy trying to get them to see eye to eye with me. Unfortunately, that is not just my friends. A lot of people who don't know me are insecure with the fact that I have changed to the point of eventually having a movie contract. When I told my mom that there existed a very real possibility that I was going to need a bodyguard, she couldn't understand me. Saying that I have changed, she said that she didn't know what to say.

The common response I got out of almost everyone was to just get a job. When I was talking about the possibility of moving into a house, I was summarily asked if I had a job. "Um, yeah, but why do I need a job when I am probably going to be able to afford the entire payment upfront?" Some people just don't get it. Again, I have come full circle and have found myself isolated with no type of emotional support. Now my finances are very slowly getting stable, but I don't have the social network that I had when I first moved to Arizona. My mother and father worry about me spending so much time on my own without any kind of social fabric. Maybe when they read this, they will understand why that happened. It wasn't something that I ordered. At the same time that I am grateful that I have found financial stability again, that all came with a high price tag – my social network. It seems that I can't have my cake and eat it, too. I am either financially stable with no social network, or I have a lot of friends and have no financial stability.

WAITING FOR THE BIG BREAK

Opening up my email, every now and then, I noticed that there were some companies reading my synopsis, but unfortunately for my interests, they were both independent companies and probably would not be able to afford my fee. A lot of independent producers pay scale wages to people like myself, and for the most part, are not signatory to The Writer's Guild, a company that I will be required to support upon the sale of my first spec screenplay to Hollywood. The Writer's Guild plays essentially the same function that a union does for workers. It guarantees what I am to be paid, whether it is for television episodes, screenplays, or if I am involved in doing rewrites on a project, how much I will be paid to do the rewrites. For each week that I am employed doing rewrites, that translates into a certain number of units of credits earned toward membership in the guild.

In the development process, the producer generally requests that the script be rewritten according to the vision that they hold for the work, or be written to attract an actor to play a certain part in the project. For example, sometimes an actor will refuse to sign an inducement agreement, attaching themselves to the script unless the script is written the way they like it. When I get to the rewrite stage, I will be trying to satisfy many masters. Once my screenplay is optioned or sold, it is technically, no longer mine. I have to be open to ideas that are suggested to me as a way to improve the screenplay.

Being a member of The Writer's Guild has a lot of important benefits. If I am a member, and a producer makes the choice not to pay me, since I would be outside the realm of small claims court, because of the amount of money at stake, something that would be put in the contract to protect me would be the use of arbitration to resolve a dispute. That is where the guild comes into play. Since I don't personally enjoy a potentially high legal bill, I would have the guild's dogs go and bark at them.

One of the ways that the guild puts producers on the edge is by saying, "You either pay up, or we are going to put your production company on a strike list[1] until you do pay up" This means that no writer can be employed by this production company. Part of the working rules in The Writer's Guild is that a guild member is not allowed to be working for a non-signatory company. A producer will also not read a screenplay that is read by agents that are not signatories to the guild, so had I sent my ex-agent in Texas my screenplay, no producer would have been able to read it, which sort of gives me an idea of how many people actually saw my work when it was submitted originally. None.

A producer who is looking for projects does not want that to be put on a strike list, because then he is not likely to get any business from projects. That makes me wonder about my literary agent. If she was not on a signatory list – and I did check – why would she recommend that I send her my script if almost all of the producers in Hollywood are bound by the working rules of The Writer's Guild? Especially since she had not yet secured a sale for my book proposal, I was not exactly in the mood to send her my screenplay. The nail in the casket was when she said it was another $300 since it was a separate contract.

Once I sell a spec[2] screenplay, my contract obligates me to become a member of The Writer's Guild, to the tune of $2,500, which is what I have to pay in dues to them

---

[1] Production companies on a strike list are not signatory to The Writer's Guild and are unable to work with those writers who have agreed to become signatory to the guild.

[2] **Spec screenplays** are written with the expectation that they will be sold or optioned at some point in the future.

every year. At that point, I have to attend an orientation meeting which went over the guild regulations and what was expected of me as a member of The Writer's Guild. If I choose not to join, then I can no longer work for signatories any longer, which is probably 99 percent of companies that produce features and every company in television. One thing about The Writer's Guild, is that you can sell to signatories without the required number of credits, but when you earn enough credits to qualify for membership, you have to join. Selling a spec screenplay to a production company immediately qualifies you for the number of credits required to join the Guild. The reason for this is common courtesy for production companies being respectful to writers.

CREDIT ISSUES

One of the things important about the Writer's Guild, is that they determine credit applicable for a picture. The Writer's Guild is who delivers official credit notification to anyone involved on the project, saying who will have their name on the ending credits. A producer cannot determine the screenplay credit. The Guild determines credit based on the number of writers on the project. In a way, it is a bad system. Because those who object to the way in which credit is assigned have to write a letter to The Writer's Guild arbitration committee saying why they should get credit. Then all the scripts are reviewed by an arbitration committee to make a final determination of writing credit. This is not just an ego issue. It is also a financial issue, because the amount of credit for the picture directly ties in with the bonus that is received for getting the movie made In addition, having your name on the picture gives you the credibility when you write your next movie.

THE OFFER

I decided to take a drastic step and blast my query letter to about 850 production and entertainment companies. Only one production company responded, telling me to go ahead and send the script. Of all things involved with writing, it is the marketing that is a big pain in the ass. And having to check out who it is that is asking you to send the material you have worked so hard on for a year, because ideas do get stolen. It is not very often that a production company would steal content from a writer's script, since that is protected under the federal copyright laws. A producer does not want to pay a million dollars for something that they could have easily gotten for twenty-five percent of that figure. Luckily, I didn't have to worry about this company too much, because this was the same company that published my friend's book about holistic medicine. For that reason, I knew that I was in good hands.

I sort of hoped that more entertainment companies would bite, so that a nice heated bidding war would get started. That actually may come later. Since I did not generate a huge response that most people experience, I figured that it would have to do with the fact that I have to be given life story rights as well as a possible role as a consultant which would provide more money to me. How many bad movies have you seen in the last year alone? I would guess that the amount would be more than fifty percent. I am not going to be suckered into constantly rewriting my screenplay according to the guidelines that a script coverage company tells me need to be in my script. Some of the information is useful. But they always find something wrong with the story, no matter what changes I make. That is when I decided that I was not changing anything else in the screenplay. I felt like I was having my story played with and that this was going to another one of those dumbed down movies. Do you ever wonder how such stupid movies get made? This is a good example of how a stupid movie gets made. For example, the script coverage company that I was using, were suggesting something to me that was totally unfounded. It would have pretty much involved rewriting 135 pages of material that was already very original in nature. I am already going to be rewriting my script to the producer's vision anyways – but not until I get a check for at least the mandated minimum as order by The Guild.

They were suggesting that my character fight the food and drug administration by suing them. Okay, bad idea. First, that has no place in my movie, second, material like that would be good advice for a sequel to the movie, if a sequel ever gets made. At this point, I am not even allowed sequel or remake rights. Because I was convinced of this, I did not even post any of that material on my website. *(Note to self: write a nasty letter to those people who made that comment that they were ruining good material.)* But if the producer decides to use those rights, make a sequel to the movie, or get a television series made off my screenplay because the movie was so successful, they have to pay me, even if I never do any work on the series since I originated the material. Talk about a nice return on investment. But if I am involved in writing the series, than I am entitled to additional compensation. The television series *M*A*S*H* was a series that was based on a very successful movie. The writers earned more in royalties than they did from the actual script sale.

FINANCIAL SECURITY

As I read in an entertainment law book, the screenwriters who wrote the original material that turned into this series made more money from royalties on the series than they actually made on the movie. That is because a series based off a successful movie gets $10,000 just for the pilot episode and then roughly $2,000 per episode, even when it goes into syndication. Just to show how lucrative a series can be, it was reported by *C.S. Daily,* an electronic magazine that I started to get in my email box that is devoted to screenwriters, that Larry David, the co-creator of the successful television series *Seinfeld* , a series that was the most watched episode for about a decade that stayed in the number one spot during the time it was running episodes. got $200 million. All those reruns paid off, because every time the show aired, a check was generated for the use of original material.

Producers, for the most part are guild signatories, meaning that they are less likely to rip you off, although there are stories about writers getting burned. Sometimes writers will pitch[1] a project to a producer and, if the writer has not yet taken the steps to protect himself by means of registering his treatment with The Writer's Guild and/or the United States Copyright Office, the producer is free to conveniently forget that this idea was verbalized by somebody else. I have not yet had to go to pitch meetings – although I will probably be inundated with them when my work sells. You can only get paid to write something when you have a produced credit for another movie. You have to be able to prove that you have the ability to put a screenplay together and know the proper format. There are many books out there for this purpose. The one that I recommend is *The Screenwriter's Bible.* That book shows you the proper style, font, and margins to use when formatting a screenplay. The reason for all this, is that one page is the equivalent of one minute.

---

[1] A **pitch** is something verbalized to a producer that the writer wants to get paid to write. It talks about the story and is told through the main character and the different character arcs that they go through.

Getting somebody to read you, when you are not a produced screenwriter and have not been published anywhere in any magazines, and have been relatively unheard of, is very difficult. People who have proved themselves time and time again are the ones that generally get work. I had been sending a lot of queries to agents, but almost all the time, they would not respond, or if they did respond, they told me that their client list is full. At the time, I had no heat. What was never told to me is that the same people that say that they want fresh material that has not been shopped all over Hollywood are the same ones that don't take new writers. It seems like it is a pain in the ass just to break into the system.

When some of these people see how lucrative my screenplay is, they are going to wish that their client lists weren't full – and that they had not made that comment to me. I better enjoy the time I have now – because by the time word gets around about my movie, everyone will be knocking on my door for something.

THE STUDIO SYSTEM

As I found out, my screenplay that I thought had a producer or director attached, was being read by about a gazillion readers at the studio. That scares me. Just to get a picture made, everybody has to be in agreement that this is a good investment for the studio and that they aren't going to end up losing money. One nay, and let the chips fall where they may. The development executive[1] has to read the script, then they have a meeting, and they have to talk the project up to everybody in the room. Based on the number of hits to my website over a forty-eight hour period, I could sort of tell where things were going. At least, I thought it was going as expected. My screenplay didn't have any flaws, but they passed on the material. The problem? There was no perceivable market.

The other thing about the studio system is that things take five times longer than you would normally expect. For example, you could have a contract signed in July and not get paid until ten months later – all the more reason to keep that day job. I know that my father is happy when he hears that, the one that seems to like routines to simply stay routines.

---

[1] Projects that go through the studio get sent to the director of development.

This is the harsh reality of the entertainment industry. I was one of those writers who was unheard of, but by the time you get this book, you may have seen my face on television doing interviews with regard to the projects that I have done. Hopefully, I will be able to conduct interviews on *The Tonight Show* and give a lot more details surrounding my story and projects.

But, I had one thing on my side. It was the unique story that nobody else could tell in any type of capacity. The only person who could tell this from the heart with the amount of passion that producers are looking for, is myself. It definitely makes it more difficult to bump somebody off the project and bring in another writer – especially when I am the only one who can give an accurate portrayal of what happened in my life.

Since I was writing this at the time when doctors were just finding out about this and there were no books that dealt with the subject of the connection of epilepsy and osteoporosis, I had that working for me. If I couldn't get people to read me because I was unheard of, it was my job to make myself heard – through an original story and 135 pages of juicy material that got passed up the food chain and got a producer and a talent management company attached to the script. I sat on my laurels – and prayed to God to make money start falling from the sky. I needed a break. Like now.

And I think after this book is out to prove that there is a market, I may end up getting that break – in the form of an option[1] which the producer would use to attract talent to the project and get a director attached. If the option is not exercised in time, then the rights to the screenplay revert back to me and I can try and set up the project somewhere else. Sometimes what can happen is that a producer can renew an option to get more time to get financing. Options which are renewed can be applicable or nonapplicable to the final purchase price.

In the process, I also took it upon myself to mail another letter to Leonardo DiCaprio's agent and let him know the status of my script and see if he was interested in playing the lead role. I figured that if he agreed to commit to my project, that my job would be much easier. If I could attract talent to my project, then I had only one problem left. I had to convince the people who would be financing the picture that this would not

---

[1] Screenplays submitted to studios to be made into movies are generally optioned at ten percent of the agreed upon sale price. Producers then have a certain amount of time to exercise the option by paying the full amount of what is due and owing for the picture.

bomb at the box office. For a first time screenwriter, that was not an easy task by any stretch of the imagination.

I dreamed of the day when I would open up my email box and get some good news. On June 16th 2004, I checked my profile at inktip.com, and found that the same company that requested my script must have passed along the information about me to one of their producers, as I did some background checking and found out that the production company is affiliated with Keats Entertainment. That was all that I needed. I needed somebody who was serious about exploiting my material to read my script and do something with it. After all, if I didn't think I would have gotten paid for my story, I certainly never would have written it. All that producer has to do now is get a studio to back him.

Studios, like banks, are the ones that provide the financing for the picture, so since they bear the risk if something doesn't become a blockbuster hit, the producer has to be the one to convince the studio executive to buy the material. It is the president of the studio that says yah or nah. If the producer can't get the movie backed by the studio, that is where it either goes to another studio, or a separate producer has to pay off the original producer for the investments and costs that they incurred trying to get the movie produced. Hollywood buys a lot of movies that just don't get made.

DEVELOPMENT HELL[1]

This is one reason why a lot of movies that are submitted to development get stuck in what is well known in Hollywood as "development hell", which is where a movie script just for one reason or another does not make it out of development and into pre-production, where the actual filming takes place. Actually, a lot of movies get stuck in development because of the amount of money that is attached to them. Another reason that movies get stuck in development is because of the fact that there is no market for the script. The next movie you go to, check the number of producer credits on the movie. There are numerous reasons for many producer credits on movies, but this is one reason. If you see more than one production company listed in the credits, chances are, the script

---

[1] **Development hell** is industry jargon for the movies that never make it out of the studio system. Very few movies get made for various reasons. For example, there are some great movies that simply can't get made because of the fact that they are not castable. Or, there is so much money against it that it stays in that studio until a producer pays the money to cover all the costs that were invested in a project.

was in more than one studio. Only a small number of movies that are submitted directly to studios are produced each year. More than 98 percent of scripts that are submitted to studios face almost instant rejection because of the storyline. That is before the script is studied for proper script formatting, proper margins, and all that wonderful stuff.

The unique side of my story had my script very quickly passed through the food chain. I think the one thing that I have working in my favor, is that I am a new writer and one with a unique story that has never been done. On the other hand, that also hurts me, because studios look at something like that as a big risk. Remember, studio execs are worried about whether or not they will still have a job at the end when all is said and done. That is why I chose to self-publish this book to prove to everyone out there the market value of my work.

New writers tend to have a different point of view since they are coming from all walks of life and from all types of environments. I knew that I would be in good hands when more than a month went by and I never heard from them. I figured that if they were going to pass on my project, they wouldn't take a month to make that decision. One day, I checked my website and saw fifteen hits on my site in one day – a very good sign that something was going to happen. But anything can happen – after all it is Hollywood. In September, after more than three months of going through impatience, I am told that there is no market for the script – three weeks after there were fifty hits to my website over one weekend.

REWRITES

Upon getting my rewrite notes back, that is when things will take a different spin. Just to get the movie made, I am going to have to rewrite it so that everyone is satisfied with the final product. I am going to be satisfying many masters once it gets to a point where I get that phone call. I will probably have notes from the producers, the actors who commit to the project, my manager, and anyone else that is involved with bringing this story to the silver screen. When the studio is satisfied, the picture is then given the green light[1] to begin filming.

---

[1] A movie that moves out of development and into pre-production, or the filming of the movie. Generally, only one person has the authority to green light a picture for pre-production.

STATUS OF WRITERS IN HOLLYWOOD

You would think that screenwriters, who are the ones who originate all this material, are given respect in Hollywood. Compare the salary that I would get with a top actor with bankable talent like Leonardo Dicaprio and you start to get the idea. At least he is probably able to negotiate for a percentage of the box office gross revenue that is received from ticket sales. As somebody who does not have a Hollywood track record, my attorney said that the best he could negotiate for me was one to five percent of the box office net profits[1]. Anybody who has worked in Hollywood knows that the true definition of net profits is *no* profits. I only know this because I spent a lot of my time educating myself about the way in which Hollywood works. This is because of all of the above the line deductions that are allowed to the studios.

According to some books that I have read, the reason that writers are treated like second-class citizens has to do with the fact that their contributions are made during development and not during the production of the movie. Once I turn in my final draft, the directors, and producers would be free to change anything they want, since they own the copyright to the script. A writer never owns the copyright to a motion picture or a television series. Actually, I own it currently, but by the time this book is printed, I probably will have transferred that copyright over to the possession of the studio, assuming that I have gotten the interest of a producer or a studio.

CREATIVE ACCOUNTING

Let me tell you something about net profit participation that you need to know. No movie in history has ever paid net profit participants. If any of them actually do, I would be surprised at that. Your contract may say that you are entitled to one to five percent of net profits, but, just from having an associate degree in accounting, I can only easily imagine all the games that studios play to avoid having to pay profit participants – sort of along the lines of what got Arthur Andersen in trouble with regard to the Enron case. You probably have heard of a lot of stories in the news about companies that manipulate the bottom line so they get a bigger bonus at the end of the year.

If you are involved with the movie industry for any length of time, that type of creative accounting is very common – and also perfectly legal from the movie studio's

---

[1] Also known as points.

point of view. After all, they are risking hundreds of millions of dollars to get a picture made, so shouldn't they have some say over how the money is spent?

Similarly, studios are allowed to use creative accounting, and take a lot of deductions to avoid paying any profit participants. So, even if a movie did well, the balance sheet will show a negative profit because of all the deductions that are allowed. Otherwise, in an otherwise perfect of perfect worlds, I would be getting fifty percent of one hundred percent of the whole. In other words, I would get fifty percent of whatever is given to the producer. The producer is supposed to divide his profits between the net profit participants. If you compare the millions of dollars that are paid to the top actors in Hollywood, let me tell you generally the percentage of the budget that a writer gets – about 1 ½ percent. So, for a movie that cost $20,000,000 to make, the writer would get $300,000 for the outright sale, which is well over the mandated minimums outlined by the guild's minimum basic agreement[1] (MBA) – but this is still not anywhere near what would be paid to any of the top actors that are involved in the success of the picture. It isn't uncommon for movies to have budgets north of one hundred million dollars. At a certain point, the studios will throw in a ceiling, which is the maximum amount of money to be paid to screenwriters for a picture to protect themselves in the case that a movie goes over budget. That is what a producer does, is supervise the production to make sure that the picture does not go over budget. Otherwise, the money comes out of the producer's pocket unless he has an eat-in[2] clause with the studio.

Actors are the ones that are the ones that are responsible for getting movies made, however, and that is why they get paid so much. And to the studios credit, they are the ones who are risking millions of dollars on something that might bomb at the box office. So, they are justified in skimming off the top, although there are a few people who would argue with me on that point.

UNEXPECTED POPULARITY

The unique story that I had not only got me closer to a movie contract, it also got me some unwanted attention. For the first time in my life, I was going to become very

---

[1] The **MBA** is the agreement that governs the minimum amount that is to be paid to writers for features, television series, or the release of the movie to secondary markets

[2] An **eat-in clause** means the studio is responsible for anything that goes over the budget.

guarded about the things that I said or the people who I associated with. Just be in a position where you might make a lot of money and you will have people start to cling to you like they know you. At a certain point, I just started to distrust everybody because I was unable to determine what the motives might be. None of us like to hear horror stories, but if you read the next chapter, you will see the method behind my madness. I never even realized that the fire I was starting was going to backfire and burn my soul and entire being.

The financial success and the possibility of a seven figure income at this point had me married to the freelance business that I had started in the form of an internet presence. The aftermath of instant success, is that once you establish a publicity factor, that doesn't go away. Not ever. You don't have the privacy that you used to have. I go over to the mall, and try to get out of my depression. I went to see a movie, and somehow, I had a bad vibe going in there.

Somehow, I feel that this is a bad move. I am sensing some bad vibes that these people know who I am. I walk past a group of people – one of them flashes a smile at me. When people around me are commenting among each other that I have a movie out, and then someone comes up and refers to me by first name – and I don't know who the hell they are – that just scared the living daylights out of me. I had lost my privacy overnight. The one thing that concerns me is somebody for one reason or another trying to dig information out of me.

My contract will specify that I am not allowed to even discuss the projects that I am working on without the express written permission of the studio – since they own the work and don't want something else to be competing with them. As a matter of fact, merely talking about who has my work or giving interviews on my own without permission is grounds for termination. I am not trying to be a jerk when I am ignore certain questions, but am more or less looking after my financial interests, not to mention trying to protect my job.

THE REALITY OF PUBLICITY

When people who I did not know started to refer to me by my first name, and I didn't know who they were, that is when I realized I had another serious problem to deal with. As a matter of fact, it was a public relations nightmare. I had to figure out how to manage the effects that my internet presence was having on my personal life – before things spiraled out of control and somebody would end up getting hurt. As a result of trying to minimize the amount of publicity I was generating, I stayed inside as much as possible – unless I was with somebody that had my best interests in consideration. There would be times that I would be talking to somebody and I would notice people walking past me who were obviously expressing interest in figuring out who I was.

One time, when I was trying to eat lunch, before I started to go to my class one Saturday afternoon, I caught out of my peripheral vision, somebody walking up behind me as if they want to get a look at my face. What they didn't know, was that I caught them out of my peripheral vision and my attention was watching them just based on how they were acting. They figured that they were going to get me in a weak position. Now, at this point, I am in a defensive position. Naturally, I look them in the eyes to get them to back down. I was trying to eat my damn lunch. The last thing that I wanted was to be heckled by somebody who maybe had it in for me. Many years ago, I would have dropped my gaze and avoided a confrontation, but now I have a lot more to lose. I stared at this person waiting for them to make the first move – and at the same time had a fist clenched out of view, behind my back. The look that I gave them, in essence, said "go about your business or things might get ugly." Luckily, they got the point and moved on about their affairs. This it the kind of stuff that I am talking about in the next chapter, when I realize that I am starting to cross the line between the life of an ordinary person and the life of a celebrity. Some people would disagree with me on this – because they don't want to acknowledge that I have changed. With God as my witness, everything that I am saying actually happened.

If you are one of those people, who is curious about somebody, the worst time to bother them is when they are trying to eat their lunch. Who would do such a thing? That is rude! Somebody who does this has no regard for how I feel. That goes for anybody with a popularity status, not just me. Some people don't care that they are being intrusive on my personal space. Little did I know how fast the word would spread about my story. I had given my business card to my advisor at school and the word got around the entire school campus to a lot of people. A lot of them were supportive and said that the work I was doing was awesome. Then there were some people who were threatened by this. The line of demarcation was about to get drawn very brutally. This would be the line of demarcation that would result in me withdrawing from school and eventually paying off all my student loans. I don't belong in school. I need to be somewhere else, but not here.

In one of my classes, I could sense a vibe of anger growing. Evidently, the word about my financial settlement with regard to my hip fracture got all over campus as well. Tell one person – and the next thing you know the entire school knows about what happened to you. It started in one of my classes when the teacher was talking about the rates of interest being low on savings accounts. Acting like a smart ass, I decide to whisper to the person next to me, who I didn't think would get the word around so quickly about my settlement, that I have a certificate of deposit that pays close to five percent interest. I didn't advertise how much I had in this CD. It could have been $1,000. It could have been $100,000.

This teacher, in a different class, let me know, in a menacing way, that she knew about the fact that I had this huge insurance settlement – without mentioning my name. When I would raise my hand in class, she would not even acknowledge me. How am I supposed to succeed in a class where the teacher doesn't respect a person because of the fact that they are different? As a result, I boycotted her class – I stopped going. The classmates who missed my presence, obviously were not getting the entire picture of what was going on. And I was in no mood to tell them. Unless they were in my shoes, they had no way of understanding what the hell I was dealing with.

If people at school were unwilling to accept me for who I was, I decided not to have anything to do with them. Either they were supportive of my efforts, or they were my permanent enemies. In about 99 percent of cases, all the people that I knew, for one reason or another, I had to stop associating with. In most cases it was because of the fact that they did not understand what it was that I was trying to accomplish. As you will read in Chapter six about the hidden price of fame, I very quickly realized how many people knew about me, as well as the fact that there were people that simply did not want to have to know about my situation. These were the people that I needed to watch out for – for my own personal security.

Either they were feeling threatened with the amount of money I was looking at, or they thought that I was the one full of shit. Who is this person saying that he fractured his hip in three places? He's just lying. With things in my life about to change, I decided to take time off from my classes, especially with a screenplay in the works. Let me state for the record right now, that I am being forthright about everything that happened. It was this incident that redefined my idea of what the definition of getting an education is in the United States was and how it related to my own situation.

MY VIEW ON EDUCATION IN AMERICA

Like I said earlier in my book, "If you are not living on the edge, you are taking up too much space." Apply that to the education that we have in the United States. We are not all the same. We have been through different life experiences. The educational system is not tolerant at all of people's differences. We have people from all walks of life that go to school, and the universities that I've been to have proven how closed-minded they are to difference. Everybody is expected to be the same, regardless of the fact that this country was founded on rugged individualism. This also told me something about education. When you go to school, all you are really learning is that you are just a cog in a big human machine. This country is putting way too many people through medical school and law school and those people, after graduation, have trouble finding jobs. Any idiot with an economics degree can answer that question! It all has to do with supply and demand. When supply exceeds demand, there are going to be a certain number of people without jobs.

And then, even if you are not performing the way you are expected to perform, you are replaced with another cog that will get the job done the way in which the job is "supposed" to get done. So much for people using their brains. With computers taking the place of jobs in a lot of fields, it is dumbing down the educational system. Okay – great – try doing that to me now. I dare anybody to try that feat. I have something that nobody can replace – life experience. Knowledge is power in the right circumstances. Some people might argue that an education is necessary in today's world to get ahead. True. But, they already established that two and two equals four. What in the hell do they need me for? Computers already add numbers up, replacing the need for a lot of the people who do work in the accounting field. In my own mind, going to school only sets me up to be replaced when I start costing too much for my employer – and then they replace me with another cog that can get the same job done for $50,000 less money. When I am protected by a union, I don't have to worry about being underpaid. I am guaranteed a certain salary increase over whatever I did last year.

Education expects that everybody is going to learn at the same pace, and to put aside their creative efforts and "educate" themselves at the expense of their minds. I'm out of here. I don't serve any purpose sitting at a desk, competing for the same jobs with thousands of other people who are fed nothing but psychobabble about how having that sheepskin is going to set them apart from people that only have a high school degree. Sometimes it is the people who have less than a college degree who are smarter than those people with a college degree. When you get your degree, you only learn one aspect of what life has to offer. I know that there are going to be people who want to put my head on a platter when I say this. But it has to be said, because every word of this is the truth.

I got an education that no school could teach. I learned about how epilepsy and osteoporosis are linked to one another. What school will ever teach people that? I will personally guarantee you that you won't find any job out there that will pay upwards of a six figure starting salary such as the job that I will eventually have with a production company – unless you happen to be working for Bill Gates.

Plenty of people got along just fine without an education. Michael J. Fox, who didn't even have a high school education, made a success of himself once he got acting jobs in Hollywood. Sure, there were bumps along the way, but who said that it was going to be easy? I have experienced the very same thing. My parents have indirectly supported my views by helping to pay some of my bills during the time that I was trying to get myself established. With the economy down the toilet, I was in no mood to go out and compete with ten thousand other people for the same job. I would submit my resume, and I wouldn't get called back. Perhaps people recognize my name and don't want to hire me because the minute I get a lucrative contract, I will be gone for good. And, to their credit, that is true. If I ever get that contract for several million dollars, I am not going to be in a mood to work for peanuts.

It is too late to turn back. I have to make my overall plan work. What I didn't realize, was that there would be indirect costs associated with my popularity, which include the loss of personal privacy, the fact that everybody knows me, and worse yet, having to worry about my personal security. When I realized what I was getting myself into, I was the one treading on thin ice with people who had no way to understand what I was doing or why I was doing it.

Every day I am reminded of how stupid the educational system is in this country. I have my diploma from the two-year school where I got my associates degree in applied science related to accounting. I look at it every day – and realize that it was just a waste of money. What is the point of getting an education when you ultimately end up doing something else besides what you were educated to do? To me, that defeats the entire purpose of getting an education. All that equates to is a career path. Even if you don't have a life story to tell like myself, I will guarantee you that what you get your degree in and where you end up working will probably have nothing to do with one another. You could end up getting an accounting degree but work in the area of publishing. The two fields have absolutely nothing to do with one another.

THE COLLEGE DROP OUT

In high school, there is such competition to get into the *right* school, because by getting into the *right* school, you will get the *right* job with the *right* amount of money and if you are lucky, you will work in one place your entire life, move up a few notches on the ladder assuming more responsibility, and build up enough in your retirement account so that you can live a decent life after your retirement party. Great – that sounds like a party that I want to go to. Sign me up! No thank you!

I don't know what it will take to get through to all the morons that keep feeding kids this endless drivel of bullshit. I read Michael J. Fox's memoir *Lucky Man*. He was diagnosed with Parkinson's disease at a young age and had been living with it his entire life, similar to the way in which I had been living with secondary osteoporosis and had not even known it. The disease had gotten in the way of his acting career and he ended up having to quit taking acting jobs and felt the need to pursue a cure to Parkinson's disease by running a nonprofit organization, The Michael J. Fox Foundation. The fact that he got Parkinson's disease was what got him involved with starting a foundation with his name dedicated to finding the cure for Parkinson's Disease. When I found out that I had severe osteoporosis due to a hip fracture, that naturally led to me changing my focus from being an accountant to being an educator – in the form of writing a feature movie script for Hollywood.

I personally enjoy flaunting the fact that I have dropped out of college after realizing that I have something that no school in this country can teach. For more reasons than one, I don't miss it and enjoy the chaos that I am about to start in this world – starting with instituting a universal language in our medical system. As you will read in Chapter 7, holistic medicine has a lot of benefits that should be taken seriously by anybody who is on medication for a chronic medical condition.

DEMOCRATIC NATIONAL CONVENTION

If there is one thing that needs to be changed in this country, it is our health care system. It is the most lucrative, inefficient, expensive industry that exists in this country. I am sure that I am not the only one that thinks this – except for those people who have an income to protect. We are the only country that does not take care of our own citizens. Over in Europe, universal health care makes sure that everyone has health coverage. While I am on the subject of health care, we are the only country that I know of that does not accept holistic medicine. While there are a number of people that disagree with me, I personally believe that by adopting holistic medicine, we will streamline the healthcare process and cut out the fat. In all of Europe, holistic medicine is used because it helps the body heal from disease.

If John Kerry ever gets elected President, one thing is for certain. The American Medical Association has to be overturned at any and all costs, even if I am the only person who will do this. After I read in an interview conducted by *The Readers Digest* that John Kerry favors holistic medicine, I went back and tried to piece together where he may have picked up that idea. This would be a first for somebody running for President of the United States – to finally have the guts, along with myself, to acknowledge the truth about the American system of medicine – to admit that there are too many inefficiencies in the system of medicine.

In terms of healthcare reform, nobody running for president ever brought up the subject of holistic medicine before I had part of my website devoted to the subject. Naturally, I pull up the statistics report on my computer and start combing through every entry over the past few months to see if there is anything that I can find. No doubt – there was a hit from the Government on my website. What is even more interesting, is that he did not access my website by doing a general search, meaning that he used keywords to find my website. He must have had my site bookmarked or had typed the name of my website into the browser. That told me one more thing. He was not the only person who was in Congress who had seen my story. Well, even if Kerry did not actually visit my site, he definitely has my vote in the November election. If a member of Congress found my story, I think I started a wildfire that is not going to be extinguished anytime soon. I like wildfires – they are difficult to contain and very difficult to put out, especially when I am the one that is fueling the fire. The only way to put out that fire would be to kill me. And I personally don't recommend that you try something like that.

Senator Kerry could have easily picked up this from his wife, but my original story provides more of a reason why holistic medicine needs to be in place in this country. He and I have the same vision, especially since his wife is from the country of Mozambique and may be able to show the American Medical Association that they are guilty of the same crimes as the tobacco companies are guilty of. By vaccinating children at a young age, they have lifelong customers and don't need to worry about where their meal is going to come from. The American Medical Association is an institution of greed that does not care about people – it only cares about money. I think that there are going to be some people who are pissed off with this idea, but a bunch of people are not going to have jobs when all is said and done. The seed of destruction has only been planted. If my story goes to show one thing, it is that synthetic medications are unhealthy for our body. After suffering an almost deadly hip fracture, I am no longer the same person that I used to be when I was going to school to get "educated." My injury has taught me one thing – that only I have the power to overthrow certain elements based on my life story.

Besides the fact that there were jerks who did not support my creative vision, the amount of money that I was making from the sale of my movie script and its underlying subsidiary rights, did not make it justifiable to get an education. I was taking a chance and for the most part, operating on my hunches that things were going to change in the next two to three months and did not want to have something conflicting with my schedule. I wanted everything open for when the big opportunity came knocking at my door.

I finished in 2001 with an associate degree in accounting, thinking that I had a skill that I could apply with numbers. My goal at this time, was to be involved with forensic accounting, especially with the scandals that were going on with Enron and Arthur Andersen. I had absolutely no clue that my services would be put to another use six months after I moved to Arizona. It all started when I was fired from Hi-Health

Supermart, the place where I was an inventory control and accounts payable clerk. (By the way, thanks for firing me – it let me know where my priorities should be in life.) As a result, I eventually took a job that would change my life – and the lives of a lot of other people.

In March 2003, I signed a contract with who I thought was a reputable agent. At this point in time, I did not have a lot of drama happening in my life, so I stayed in school – the only thing I was doing outside of school was working on my book. After I was injured and off of my crutches, I enrolled in school, trying to get everything back on track. Even I was in denial that things were eventually going to be forever different. At the time, I was writing my book, before it was seized by the sheriff's department, for the purpose of using it as evidence in their investigation. I didn't feel right trying to get another agent to take this, because the value already dropped significantly because of where it landed.

I didn't know how I was going to get the ball rolling again. It seemed like I had a knife put through my chest. As I would find out, this was exactly what I needed to give myself publicity for my story. I figured it wouldn't hurt my case, either, to put this on my website and get some people feeling sorry for my circumstances and give me benefit of the doubt. Another six months of work, and I was for the most part, going to be a produced screenwriter – or at least getting paid to have my screenplay optioned for production.

In May 2004, I took a break from classes figuring that big changes were bound to happen. For the first time in a long time, I was starting to get scared about whether or not I was making the right decision. I was told that there was a two month reading period, but I had not heard back whether or not things were going to be proceeding as I had planned them to proceed. I knew that there had to be some reason things were taking this long. After looking at the time line, I was completely lost and when I got an email saying that my script was under review, I didn't have any idea who would have been reading it. As I scanned inktip, I found that my script must be in good hands. Because of the fact that my script came to my talent manager, after the people at the production company had read it and done coverage on it, that had more weight than if I had sent it in directly to them without someone else reading it first.

The last thing that I wanted to do is return to school and have to deal with the "troublemaker teacher." You know who you are, too, if you are the person I am referring to.– you were the one that taught income taxation for businesses and would not recognize me when I had my hand raised. Something is wrong with you – go get a psychoanalysis.

There is only so much a person can take before they decide to take matters into their own hands – which is what I did when I made the decision to drop out of school, pay off my student loans, and never, ever deal with the business world again. There is definitely more to life than accounting. That was sage advice that was given to me when I was working at a health food store as an accounts payable clerk. At the time, however, I didn't know that I would end up being a teacher.

The people that said that to me were right on their own accord. It was going to take the right set of circumstances to make me take a completely different look at life. Not only that, but I felt that I had nothing to contribute to the world by becoming "educated". The only education comes in the form of designing you so that you are just like any other piece of human machinery operating the big human machine. What is the point of competing for the same job that ten or twenty thousand other cogs are trying to get? Try to sell my mother the logic in my thinking. She will never get it – as will a majority of people who have yet to realize the potential that they can offer to the world sans a liberal arts education. I think it all has to do with money. You notice how school guidance counselors never suggest moving to Hollywood or working in show business? I think that they have hidden agendas and get kickbacks from every unfortunate soul that they get signed up at other colleges.

Education does have a place – it separates the less fortunate members of society from those people who are on the higher rungs of the stratosphere. Of course, to get to that level of thinking, you have to get an education first – only then will you realize that you can make a decent living sans a liberal arts education. I got my education. You know what I found out? Despite the fact that I had a compelling story, the entertainment company passed on my script, saying that there was no market for it. I was devasted. Then I realized what I had to do. I had to find a way to get my script in front of actors who would agree to do this script. Well, it's back to square one.

Chapter 6

The Hidden Price of Fame

Before I started down this road, I never realized the impact that I was going to have on people. I would be going from a no-name person to a celebrity in a matter of a few months. As I walked to the movies after my foundation was ripped from underneath me in January of 2004, I was getting smiles from people that I did not even know. It just hit me. I was well-known by a lot of people – who were seeing my face appear on their desktops via my internet presence. I was getting bad vibes from this situation – and wondered if I made the right decision to completely change my path in life from being a numbers-oriented person to someone who would have a unique voice as it relates to secondary osteoporosis.

I walk outside and go to see a movie. At the time, I was without representation and had no idea of things would ever be the same. Some people outside say the words "He has a movie out." At the same time, they are looking in my direction. My gaze drops. Somebody walks up behind me and says "Eliot?". The only thing I could do is to not acknowledge the comment. Pronouncing my name incorrectly didn't score any points with me either. I figured, if I did acknowledge who I was, the flood gates would be opened. Shit. I think that I need to have a bodyguard. At this point in my life, I was flat broke and wouldn't even be able to afford a bodyguard. I will guarantee you, and mark my words, that when I hit the big leagues, I will have a bodyguard when I go out to a public event.

This situation was just beginning to get to be too much for me. Now, I have to not only worry about my own security, but that of my immediate family, who I am pressuring to help finance this dream.

I was told when I first started writing this book, to talk more about my own experiences. So, I going to use that as an excuse and throw caution to the winds. I hope nobody out there gets offended at what I have to say in this chapter.

RECOGNITION

In order to understand the direction that I am going with this chapter, you have to understand what my motives were. When I was in high school, and having to spend a lot of time sleeping in the nurse's office as a result of having a seizure, I felt like nobody knew who I was. With no connections at all, the only way to make friends was to be in a sport. In my freshman year, my parents decided that it would be best if I were on the swim team. While that was an individual sport, it also meant that I would have to jump into freezing cold water every day after my classes. Not only that, but that would mean that everything else in my life had to take a back seat. Even on the holidays, it was expected that I show up for practice. Never mind the fact that holidays are the time that you are supposed to spend with your family. There was a coach that preceded the one that I had at the time that took that a whole lot further, and made everyone come in on a Saturday as well. I was ready to blow up. If this is what I had to do to get recognition, I was at a loss for what to say or do.

I always told my father, that all I want is for somebody to understand who I am. I think that this was the worst four years of my life. As I looked around, everybody had a letter on their school jacket, which meant that they were on the varsity level on the sport that they participated in. Because I was sheltered in my own little microcosm, I was convinced that this was the only way in which to get attention to myself. My parents had bought a school jacket for me when I was in my sophomore year of high school, and I stopped wearing it. I figured, if I have to look different, being the only person without a varsity letter on their jacket, I am just not going to wear it at all. At one point, someone said that they had a way to get me a letter for my school jacket and I declined the offer. If I was going to get the recognition, I wanted to earn it legitimately. I was not going to look like some sort of wannabe.

If there was one thing about being part of the swim team, it would tell me where I needed to be in comparison to where I was. I just didn't know about it yet. As a result of my intolerance of cold temperatures, I dropped off the team and focused on my part time job working at a library, building up some savings.

I had enough. I told my father, that this was all *their* idea and I was just going to drop off the team. The next thing I have to deal with is a father who does not like people to be able to think for themselves and make their own decisions. He said "I am your father and you are staying on the team!" That never lasted. I knew exactly what was going on. He wanted to be able to live vicariously through me being on the swim team. His father had told him that he should be a social worker and he did that. Thinking that he was going to be able to manipulate my mind in the same way, he decided "father knows best."

Eventually, the same thing extended over into my freelance business, when I found out that while I had gained my own sense of recognition, my father did not support the idea of me being self-employed. My parents started to meddle in my affairs, when they had no business doing that. Fearful of the fact that the amount of money I would generate would far surpass any money that he had ever seen, he did everything that he could to make my life difficult. When I told him that my job prospects were growing thin, he told me to tear my website down and see what happens. What he really wanted to do was use mind control and thought by not supporting my visions in any way that I would just surrender peacefully and get a job.

My brother had dropped out of everything that he had been on. As a result, he was not planning on making life easy for me. Maybe my brother had a good reason for dropping out of a lot of things. I know I am going to get some crackpot telling me that as long as I am under their roof, I have to do what my parents say. My response to that is, it is my body, I don't feel comfortable jumping in the cold water every day after school, and I am through with this, and I am *not* going to allow my father the pleasure to live vicariously through one of his own sick fantasies!"

Ten years later, when I moved to Arizona, where you could probably fry an egg on the sidewalk in the right circumstances, I found how much the real world told me about my family. Not only was the warmer weather better for me, it also was going to lead to prove dangerous as I had a condition which was referred to as bone softening. I never thought in a million years that this would give me what I had been looking for my entire life, which was attention, not to mention the ability to speak through my experiences.

By taking a job that I normally would not take, I would discover something that was putting several million people in danger. At the time, the medical profession did not know that epilepsy was linked to secondary osteoporosis – actually, around the time that I had my surgery, there were research studies done in other countries with regard to the effect that some of the earlier medications have on bone density.

Bone density tests and bone scans were not routinely done on young individuals because the statistical data showed that most of the victims of osteoporosis are those over fifty years of age and female. From a statistical standpoint, there are far more women that develop osteoporosis than men. Of the ten million people who already have osteoporosis, eight million are female.

What I didn't know, was that I would have the exact polar extreme of the problem that I was originally dealing with. Instead of nobody knowing me, everybody would know me. Because of this, I felt like I had the same problem as nobody knowing me. As a result, I felt that I had to protect myself from people who would try and possibly attach themselves to me. I was unprepared for what I was getting myself into. Rushing into getting my story out to the public and putting all my energy into my freelance business, I did not know that not only was I going to find that I had a lot of interest in my story, but that my problems were only beginning when I could not get representation for my work. Compounded by this, with my story all over the globe, I was blacklisted when I tried to apply for jobs. The next line of demarcation was *again* about to get drawn and relatively quick.

Despite the fact that my parents were supporting me, I issued an ultimatum to them. At this point, I had to have people behind me either one hundred percent or not at all. I did not want to have anybody on the fence. And I could tell based on my parents actions that they were not really there for me, regardless of what they said. All they said amounted to lip service.

With my entire family hanging on by a shoe string, I felt I had no choice but to self-publish my book. My car was close to being repossessed, my mother had been paying my rent for three years straight, and I had been out of work for more than a year, which was becoming more and more difficult to explain in a job interview when I was asked to tell about myself. I simply cannot wait around any longer for a phone call from an agent saying that they want to publish my story. After all, I ran into the same thing with my movie that I did when I first tried to get an agent to represent my book. They did not see any viable market that could be exploited.

MY FIRST TIME IN CALIFORNIA

On March 19th 2004, I had arrived in Beverly Hills, California to meet with my attorney and a producer. As I am driving along on a street in California, out of nowhere, I hear somebody exclaim "Nerd!" loud enough so that I would hear it. I had no idea what this would have been referring to. I thought that perhaps it had to be from my out of state license plates, which spelled "outsider" in a very loud way. As I would find out months later, it was exactly what I was thinking originally. It had nothing to do with my out of state plates.

As I walked into the coffee shop where I was going to meet this producer to go over my script, I am given a little bit too much attention, which I do a very good job of ignoring as I start making notes before talking with the producer. Every now and then, I catch someone walking within close proximity of the table to find out what the hell all the fuss is about. A lot of times, as people smile at me, I recognize them, but don't drop any hints about my identity. It seems that everywhere I go, somebody seems to recognize me. I didn't think that *this* many people knew me. What did I get myself into? As I wait at the table, checking my watch every five minutes, for the producer that I am supposed to meet to do a critique of my script, a mob of people start congregating around the area that I am sitting. I ignore the activity and make notes while I wait for the producer to arrive. He spends two hours going over everything, line by line, saying what I need to put into my script to engage the audience. Once that wraps up, I prepare to head to my next appointment.

As I depart from Los Angeles and head to Santa Monica, my attorney recognizes me before I recognize him. Dressed in a suit and tie, I am being very loud in the way that I am different. I want to put an impression on him and figured it wouldn't be in my best interest to be wearing street clothes. For a long time, he is the only ally who is looking out for my interests.

He and I sit down in a corner and he tells me about himself and his background in the entertainment industry. I am contemplating how difficult it is going to be for me to get a sale for my screenplay. Telling me that securing a sale for my spec screenplay is low because of the demands of the market, I start to wonder how I am going to get a sale. I had written well over one hundred letters with no response. He then asks me if I know of anybody who would likely be interested in playing the different parts in my screenplay. At this point, I had not given that a moment's thought.

He tells me that the only way he can be of assistance is if I can get a bona fide offer from a third party. Asking me if I have any more questions, I can't think of any that I have at the time and we depart and go our separate ways after I sign the retainer agreement – and pay him $500

I always thought that producers like to create their own team without the input from the writer. I had thought momentarily about Leonardo DiCaprio playing the lead role, and went back to watch his movie, *Catch Me If You Can,* and asked myself if this would really be workable. Would this movie be similar in any way to my unproduced screenplay? Thinking about this for a long period of time, it was originally based on a best-selling book. I realized what I needed to do to get the ball rolling. It was not going to be easy, but it was my last resort. It would definitely buy me some time so that I could think about how to get the copyright to my screenplay sold and really get my story released to the world.

FRIENDS APLENTY

Anyone who has made a fortune for themselves could probably tell you first hand how many times people have acted like they are their friend – for the sole purpose of getting their money in some fashion. Ask anybody who has struck it rich winning the lottery, and all of a sudden, it is as if everybody knows them. This is something that terrifies me – and it is something that I am watching for. Now that I have a lot more to lose, for the first time in my life, I have to look for ways to prevent an attack coming out of left field. My mother says to me, "Until your name is a household name, you have nothing to worry about, Eliot." My mother is living in a sheltered, closed-off world. Not all of us live that way. There is no way to make her understand what I deal with everyday. What she was attempting to say was that she was unsupportive of my vision. Trying to get my mother to understand the problems that I had to deal with when I was in public was about as difficult as it would be for a white person to understand why somebody of color would be profiled by police. Until my mother spends a few days in my shoes, she will have no way to understand what I have to deal with on a daily basis.

Believe me, when I all of a sudden hit the trades, and the word gets around that my screenplay has sold for a high six to low seven figure amount, and am seen on television doing interviews, people are going to automatically assume it is because I have a shitload of money somewhere.

Go ahead and smash that theory, mother. That person who tried to run me over with their car when I was crossing the street was doing it with the intention of killing me – because my bodyguard had to physically shove me out of the way. My mother lives in her own little microcosm and has this habit of believing that everybody should live the way she lives. I can understand why she is pressuring me to get a job. What she does not understand is the fact that when an employer asks me to tell about myself, it becomes difficult to separate my personality from my celebrity lifestyle.

She defends this by saying that she is stating her opinion. Fine – but you weren't there to see it. I don't care if people judge me based on what I say. But on the flip side, you don't know what that person's situation really is and what they are dealing with. Everybody judges everyone else in some way or another. I could understand if they were there and had an observation, but I am telling them what my concern is and they are dismissing me completely as though what I am saying is bullshit. Their usual response is "Well, I was not there, Eliot. So I don't know what happened." What they are really conveying to me, is that they don't care about what I am sharing with them and don't even want to know.

I am convinced that the person saw me, since it was daylight outside, and decided to put a scare into me. Having to deal with an enemy that you don't know exists is worse than the terrorism that hit our country on September 11, 2001. Anytime I would walk into a convenience store to purchase something, I would hear the words mentioned such as "snob" or "nerd". These people don't know me for Christ's sake. The people that say this have no idea the type of person that I am on the inside. We all secretly judge people who we don't know or like, but the people that verbalize it, from my experience do this because they want you to know for one reason or another.

It is easier for people to be negative about other people than it is for them to understand *why* they are doing something. The people who say that I am a "nerd" or a "snob" don't have any idea what my motives are, what my background is, or why I am hell bent on trying to persevere. As a matter of fact, as I entered this store, I knew that there was danger lurking about. Since I am so tuned into my environment these days, I can tell when trouble might be around the corner.

When I tried to explain these situations to my parents, my mother was so insensitive that she had no way of being able to understand what was going on. Even wearing a baseball cap to shield my identity and dark glasses, I am picked out at the airport after I get back to Arizona one day. Not too long after I got home, I had blood pouring out of my forehead – because somebody had the nerve to punch me in the face –

and had to get ten stitches put in my forehead. I was somehow, followed home from the airport by somebody who wanted to harm me. For the next week, I had migraines around the clock.

That example is a good illustration of the misunderstanding and anger that a lot of people have when it comes to the problems that celebrities and well-known individuals have in terms of dealing with society. Being thrown into the fire with no way to figure out what I needed to do, I was a little bit vulnerable. Nobody else in my family had the notoriety that I had.

I am sure that my mother told my father about this incident, all after expressing the concern to her about this security flaw – what she probably didn't tell my father was that I was directly expressing my concerns to her while in Milwaukee. All my concerns fell on deaf ears. Now I completely understand what happened during September 11, 2001 when somebody faxed a memo to the FBI relaying information about an increased attendance at flight schools consisting of those of Saudi Arabian descent interested in getting an FAA license to fly airplanes. Whoever read that memo probably thought that somebody had a racial prejudice against those with Middle Eastern descent. That fell on deaf ears – and then 3,000 people died about a month later after four planes were hijacked and flown into the World Trade Center Towers. Sometimes the consequence for not listening to a serious concern is the loss of human life or personal injury.

Applying that to my situation, if it happens once, it is more than likely bound to happen a second time. And a third time. Most of the time, this is done to inflict terror into a person. Sometimes, it is done with the intent to harm. It doesn't take a psychology degree to tell you that people who do this do it usually with the sole intention of inflicting terror and making the person feel like they are in jail.

When I am going to end up having to pay $30,000 a month for security around the clock, which will be a drain on my pocketbook, you can understand why I get ticked off. I have to pay for the fact that I am different from most people in this country. Next time you go to that Rolling Stones concert or any concert, and complain about the ticket price being so high, maybe you will understand why it is so expensive to get tickets. Anything you buy in this country, whether it is a concert ticket or something in a retail store, 25 percent of the cost of what you pay for is for increased security – because extra security is needed to either prevent people from stealing merchandise, or prevent people from heckling those who have a lot of money. When you have a lot of money, people are after you for one reason or another thinking that they might be able to get something for nothing. That is exactly why somebody, who I assume was stalking me, tried to open the door to my apartment one day while I was sitting inside. Lucky for them, the door was locked – otherwise my hands would have been on the pepper spray that I bought for protection from that bullshit.

Even before I had a contract for my movie, it seemed that people were after my financial status. When I went to get a document notarized at the bank for my trade name, I was asked if the business had a bank account. That would be a natural question to ask, but I don't think that would be asked unless there was some indication that the person might have a lot of money coming to them.

If I thought that I had problems getting people to understand my plight, now I had a new problem. Because I was too well known, I had to distrust everyone, because they all wanted to be my friend, for one reason. Maybe I am better off just being alone. Of course, that is not what I had intended to happen. The more I go into a store to get something and hear someone talking smack about me behind my back, the more I realize that I am completely isolated. My parents, who were telling me to get a job had no idea that I couldn't get a job, even if I wanted to. Almost everyone knew about me and figured that I would be a disruption to the work environment.

The isolation would all start in the job interview, when I would be asked to tell about myself. Since at the time, I had been unemployed for at least a year, I had to explain that I had fractured my hip in three places only six months after moving to Arizona. Why do I disclose this information? Because first, I would much rather this information come out in the interview process so the company can't claim that I wrongly withheld information from them, and second, my freelance business, Trimberger Enterprises, is listed on my resume to fill in the gaps that I have in my employment.

Again, I didn't have any idea that this was going to happen and be such a problem. That must mean that more people know about me than I am currently aware of. But if I have my way, I won't ever have to worry about job security again in my life.

CELEBRITY WITHOUT PAY

When I was growing up in Milwaukee, it seemed that I could not walk around without somebody looking at me like they knew me. I don't mean to sound cute, but, I was a little too handsome – and was quickly growing tired of the attention. I guess that is what resulted in my move to Phoenix, Arizona. On the east side of Milwaukee, there were bars all over the place, especially in the inner city . Tired of dealing with all of the attention, I started to not acknowledge people as I walked by them on the sidewalk, dropping my gaze as I walked past them. Why are they staring at me? I didn't do anything to deserve all the attention.

I would often go so far as to cross to the other side of the street if I saw trouble up ahead. All that resulted was that people followed me more and more. That is when I started to carry an attitude with me. I never smiled – and people could not understand why that was.

I remember complaining to my parents about the fact that I couldn't go about my own business without somebody paying attention to me. People looked at me, as I carried my notebook computer – one of the only people who carried a notebook computer to school – and the name that came to a person's mind was "nerd". I actually carried the computer to class because at times, I couldn't even read my own handwriting or remember what was said. I typed faster than I was able to write the notes down. As I moved to Phoenix, I blended into a city with three million people and was able to start my life from scratch. It was by doing this, that I learned more about myself. It would also end up where people who I had never met started to casually refer to me by my first name – because of my internet presence and the picture of myself that I have on my website. My original story didn't help too much in this regard, either.

THE VERDICT

In December 2002, I got a call from my attorney's office that was dealing with my worker's compensation claim, who had told me to come down and sign some papers to pick up the worker's compensation check that had been arranged. My attorney, with enough negotiating power, had been able to get a very healthy settlement as a result of the triple fracture of my left femoral hip. As I drove to the bank to deposit the check, I was not expecting that someone would remember me from that day forward. The teller examined the check and wanted to see who was giving me a check for thousands and thousands and thousands of dollars. Why would I be getting compensated when I don't have any visible injuries? If that were not the beginning of my problems, I would really see it hit close to home in the next year – when I started to serve my sentence doing hard time.

When I had started working at an insurance company as a temporary employee, I found that a lot of people had found out about how much my worker's compensation settlement was – without me even mentioning a word. And they all wanted to be my "friends". What they didn't know, is that I took pretty much the entire sum of what I deposited and wired it to an out of state bank located all the way on the other side of the country because I knew that I might have a problem on my hands if I didn't do that. I know who my true friends are. They are the ones that want to be with me because they want to know more about who I am as a person, not because of the fact that I have a one million dollar paycheck or a healthy settlement from worker's compensation.

LIFE CONTAINED TO MY OWN JAIL

Everyone who thinks that being popular is the one thing that will solve their problems, listen closely. Popularity separates you from society because you are different. A lot of celebrities who have a publicity factor in one way or another, compare being a celebrity in a lot of cases, to living in jail. I don't blame them. I happen to be one of those celebrities.

In a way, you could compare living as a celebrity as like life living in jail with no freedom. My only way of getting some quality of life was from my view out of the second story window of my apartment, located in the living room as I typed at the computer. Recently, I would notice that the house across from me would turn their light on outside the house. It was never like that before. I never noticed that the light was on before. Before, I would have a view of the city lights over by the mountains. Because I did not want to give anybody the satisfaction of thinking that they were getting my attention, I adjusted my blinds so I would not see their blinding light any longer. Shit - now, I can't see outside. I can't wait until I am able to move out of this place and pay for personal security. I am willing to compensate for the fact that some people want to make my life miserable.

The only time that I would emerge from my apartment would be to fetch the mail. I spent a lot of time inside my house. There would be times when I went outside that I noticed somebody sitting on the stairway that clearly had no reason to be there. I later would find out why they were sitting there. I was not about to blow my cover and let them know that I knew about this.

At 3:30 one Friday morning, I walk over to my computer to check the email like I normally do. Ever since I have been feeling the pressure of the celebrity lifestyle, I have been taking security measures that I would not normally take. My blinds are shut at all times and the lights in my apartment are always off. The only sources of light come from my television, which I use to periodically play rock music so I don't go crazy and from my computer, which is on all the time as I sit in front of it – with my eyes burning up as though they are on fire. The only time I will generally go out is at night when I cannot be seen as easily. The last thing that I want to have happen is to be cornered by somebody who wants attention, which happens fairly often, especially at the grocery store. Strangers will look at me, or as I am proceeding to pay for my groceries and scan them, somebody will try and cut me off because they want me to notice them.

I hear some sort of party next door to me. I later determine that it is not what I think. I think that I am hearing a bunch of people somewhere getting drunk and reveling inside their apartment next door to me. The next thing I hear is somebody blocking access to me, somehow. I hear a female's voice, hell-bent on gaining access to me, while I am locked in my apartment and not planning on coming out while all of this is going on. This must have been the same person who yelled my name loudly while I sat in my apartment on the couch typing on my laptop computer. I knew that this was a person to avoid at all costs.

This person had the courage to supposedly keep the peace by blocking access to me, saying "He doesn't want to be bothered." Kudos to whomever was responsible for that executive decision. If that doesn't spell out for those doubting me as to why I need to have a bodyguard, and why I will have a bodyguard when I go on tour in public, I don't know how else to convince you. I have never met this person who was trying to gain access to me, but I think I know who they are. I think that they followed me to Arizona all the way from Wisconsin. Ever since the statistics count from that area of the country dropped significantly, it makes me draw that conclusion. A good majority of hits coming to my website are in either California or Arizona or the east coast. Whoever they are, they only want to associate with me for one reason that I can think of. Forget it – it ain't going to happen.

There would be a number of times that I would hear this same female voice, upset at not being able to gain access to me. Luckily, I had dumped my landline service on my phone two years earlier. I had read a report in the newspaper that a number of people, upset with their local phone company, were ditching landline service and switching to having only a cell phone. As a way to save money, and get free long distance, I decided to do that – having no idea how much that was going to save my ass. Because if I did not do that, chances are, she would have my phone number by calling 411.

With no way to "find" me, I was, in my own way, protected from the craziness that was erupting in the entire world. Having an unlisted number is one of the first lines of defense against fanatics and crazy people that may want to do harm to me. Heaven forbid I ever decide to buy a weapon and use it against someone. Gee, I wouldn't be understood at all for doing that. Being on the second floor in my apartment complex is a major asset as well. That way, people can't just come up and knock on my window to get my attention. Hopefully I will be out of this place very soon and be in a more secure setting.

During the time that I was trying to get a contract in place for *Euphoria*, I was doing a lot of clerical work for different temporary agencies. For the first time, I was walking into the work environment looking at people and making a mental note as to who might be a troublemaker. Call that whatever you want. I call it "personal assessment" and "self-defense". It seemed like every place that I went, I was immediately recognized. I can tell just by a person's reaction to my presence, just based on their nonverbal language, whether or not they have read my story – and how they regard me as a person. The big tip off at one of these companies was when I was asked what my last name was. Okay, I am not going to lie about my last name, but that is not a typical question that I am asked by a complete stranger. But, for those people that asked me what my last name was, I have my eye on you for problems in the near future. If I had not dropped my landline phone, I would have lied about my last name. My own personal motto – "When attacked, attack back." Think about that for awhile. If you have a certain level of sanity, then you have nothing to worry about.

One afternoon, I received a call from a temporary agency that I was registered with, saying that they had found a job for me. The catch? The client wanted to meet me *before* the assignment started. I immediately got suspicious and inquired for further information – the exact words were that they "wanted to see what I looked like." I had never been asked something like this before and I immediately declined the assignment, despite the fact that I knew that I needed the money. I knew that I would be getting myself into something that resembled a trap.

Very slowly over the course of a two year period, it became apparent that I had a public relations nightmare erupting and had to do something about the problem – and that is when I put tracking information on my site so I can see anything that I want in regard to who, what, when, where, why and how my site is accessed. For those of you who think you are invisible, you are not – so don't delude yourself to that fact. If I want to get information about who exactly is accessing my site, I have every way underneath the sun to do that. Welcome to the brave new world. Eventually, I am going to have to pay a full time personal relations agency to handle my personal affairs, especially when it gets to a point where I am getting different requests for interviews. The other reason I will have to do this, is because I have no intention whatsoever of putting my phone number on my web site because I don't want the general public to have easy access to me.

If you think about it, I am on the other side of that fence all the time. Whenever I write to an actor or actress, I have to contact them through their agencies, which I get by calling the screen actor's guild. Just like the agent is the gatekeeper for actors, the personal relations agency is my gatekeeper.

THE CITY WITH BRIGHT RED ROCKS

If you have never been to Sedona, you should go and see it for yourself – and be sure to bring your camera. It is the only place in Arizona where you will see red rock formations lining you on all sides. On one of the pages of my website, I have a bunch of pictures that were taken of Sedona. If one thing is certain, it is that Sedona is unique in every possible way. Just thirty miles outside of Flagstaff, red rocks line the entire landscape. I was told on one of the tours that I went on that the rocks are red because of the iron content contained within the rocks. Whatever it is, the rocks make the landscape look real scenic. If I had the money, I would try and get up to Sedona a little more often than I am able to. For my birthday, my parents decided that they would take me up north to Sedona, a scenic part of the Arizona where there are a lot of red rocks. You would think that getting out of Phoenix and driving 200 miles north would get me out of the spotlight. It did not get me out of the spotlight. There would be some people that would recognize me up there as well.

I needed to get out of Phoenix because of the publicity. However, it seemed that I was recognized up in Sedona. While my picture was being taken, somebody noticed me, even with dark glasses on. I knew that they were not sinister; they just wanted to get a picture of me. Whether they saw my picture on the internet is unknown, but this is the first time in my life that my name had some meaning attached to it – and the first time that somebody requested to get a picture of me. I had no fear of this person. From checking the stat records on my website, I could tell that a lot of hits were coming from the Arizona/California area of the United States, so that wouldn't surprise me. If there is fear in the air, my personal radar will surely pick it up. I have a thing with being able to pick up a person's vibes – something my mother does not understand about me.

Somehow, as my picture is being taken, I find it to be unusual for that to happen. This is the first time a stranger has requested to take my picture – and the first time that I realize how far reaching my story might be.

PYROMANIAC

With the unique nature of my story, and the execution of the plot that I put together for my movie script, it only took me a little bit of time to get a nice forest fire brewing. Looking at my website detail, there was one time that I had about fifty hits to my site in one weekend alone. I knew that something had to be up when I saw this deviation on my website radar software that I installed. You can look at the amount of heat generated on me as an advantage. If everyone is talking about me, then somewhere, there are going to be bidding wars over this hot property that has Hollywood set on fire as I write this paragraph. If 98 percent of the scripts that are submitted to studios are facing instant rejection, then I must be doing something right. However, there is a flip side to that coin.

With the increased exposure comes extra security concerns that I have to worry about on a more personal level. It is the one thing that has kept me out of a lot of public places unless accompanied by a trusted companion, especially shopping malls where there are huge crowds of people. While there may be safety in numbers, that also applies to those who may want to harm me. Believe me. It already happened once to the tune of $600.

Even when I go to movies back home, I am almost immediately recognized by those around me who ask "Who is that guy?". Having cognizance that this happens a lot, especially in movie theaters, I make sure that I am always accompanied by at least two people preferably more at all times.

THE MEANING OF CELEBRITY

According to Merriam Webster's dictionary, the definition of "celebrity" is defined as "the state of being celebrated". More specifically, this means that a celebrity is "widely known and often distinguished, renowned, noted, famous, illustrious, notorious", a definition which is taken from the adverb "celebrated". Some people in my family would argue with me, but I had been creating a name for myself and they had no way to understand it because they were not experiencing it first-hand. For once, I had things working for me. While other writers toil for several years to get quality work into a producer's hands, I did the same in under nine months, at least when it came to actually writing the entire screenplay.   I would eventually be breaking down barriers all over the place and for the first time in my life, I was experiencing a situation where my parents were unable to understand me when I said that I might need a bodyguard because of my unique status. The following example drives the point home.

WHEN THE HELL ARE YOU GOING TO GET A JOB?

My mother had no way to understand the unique and very serious problem that I was facing. I was dependent on my temporary agency to find me work, because no company in their right mind was going to hire me. While the fact that my story getting around was good, it was preventing me from being able to produce income. A lot of employers don't want to go through the trouble of hiring a person who is going to leave six months later after they have a lucrative contract from Hollywood. What these same people don't know is that when there are tens of millions of dollars at stake, nothing happens overnight. As a result, due to the public ignorance, I was pretty much unemployed for the better part of 2004. If you can't understand that, think about it from this aspect. The reason this book got self-published, is because it is difficult to find someone to be an agent for this book, because they don't see a viable market. If the market is not obvious, then regardless of how real the events are, it is not commercial and I am on my own. My goals are to get this book picked up by an agent once the sales hit a certain level.

This was another aspect of being well-known that was terrifying the hell out of me. My blueprint for financial security very well could backfire and blow my face into a thousand little itsy-bitsy pieces. In a way, I wish that would happen, because I am starting to get sick of having my dreams supported by my own parents – and sick of having to hear about it on a weekly basis. Some people are afraid of the phone with good reason. Think about that if you think you are going to change the world overnight. What I failed to realize, is that it would, outside of my own control, involve people in my family that did not want to be involved. What I saw as positive change would be reflected right back onto me by my family. Unable to accept my choice of "occupation" they certainly made my life difficult. I am sure they will be there when the money is in the bank.

Even though my mom had been paying my rent ever since I moved to Arizona, it was getting to a point where I was starting to resent talking to my own parents. Every time we talked, the subject of money came up – or the lack thereof. I had been sacrificing my health doing everything that I possibly could do to get my work in the hands of as many producers as possible. With Go Girl Media, a producer in association with Keats Entertainment out of the picture, saying that there was no market for my work, I was completely stuck. No employer was going to hire me. Because the word about my story had burned all around the country at a record pace, before I had a contract, I ended up shooting my damn foot off. I was not able to run anywhere. Going out to the mailbox was something I started to dread. When I hear that someone says that there is an inadequate market, I want to take their head and dump it in a bucket of ice cold water. Where the hell have they been living for the last twenty years? The numbers are right there! If you want to see those numbers, check out my website where I have posted Powerpoint slides with respect to a marketing plan. It clearly says that there is a market. I am not going to go to the trouble and explain that in my book. The reason that this book is put into print-on-demand, is because I am attempting to prove that there is a commercial market so I can get the ball rolling.

A HIGH CONCEPT IDEA

The whole issue, is that it is easy for studios to reject something that is high concept, especially something that has never been done, and it is even more easy for a studio to pass on a writer that does not have a produced credit under their belt. When putting a proposal together, they tell you to find competing and similar titles that have been done recently. I have even had people ask me if that is really a true story about what happened to me. Maybe there is no market, because people need to be educated about that which they do not yet know.

I think that I will use that as a negotiating point with my attorney so I can work as a consultant when my story finally does sell. And it will – I will make sure of that. I didn't come this far to be told that there is no market for my work. I have the information on my website to show that there is a market based on the fact that I have hits from all over the country.

The other thing that really put me in an enviable position with a lot of people that I don't know is the amount of money that I could make compared to most people. What they didn't know, is that I am pretty much cash poor at the moment. I just wish that I was able to see some form of light at the end of the tunnel. The one thing that I had that nobody could replicate, was a marketing nightmare from the financiers point of view. Who wants to take a risk on something that might ultimately flop at the box office? I figured that I had to be more aggressive with the people visiting my website. My story was not enough. I had added on another page which had marketing plans for why this movie should be made, along with statistics from what I consider to be reliable sources.

THE PRINCESS OF WALES

It was a Sunday night sometime in 1998. The television stations were all reporting one thing: the death of the Princess Diana of Wales, England. Princess Diana died from a car accident after the car she was in lost control in a tunnel in Paris, France. I was frustrated because every television station was airing the facts surrounding her death. At the time, I was thinking, "Okay, what is the big deal? This is a half world away from me.

It does not affect me on any level. What could be so important about this person that every television station has to cancel my television programs?"

While these questions that I pose sound crass and to the point, I am trying to illustrate what goes through most people's heads when they relate to my unique problems. Because I had no idea what it was like to be in the spotlight for any length of time, I couldn't understand what would make somebody go to any length to do something that would result in a famous person getting murdered. It was because of the fact that, at the time, I was not a celebrity that I was unable to show empathy for those who were most at risk – until now.

At this point in my life, I did not know why there was such a media circus surrounding her. I would later find out that the paparazzi[1], who will stop at nothing to get a good photo shoot of somebody to put on the tabloid of the papers, had been partly to blame for the accident that resulted. Although, depending on who you talk to, you can get conflicting stories on that, so I could very well be wrong. Since I don't have all the facts, I am just going to leave it at that.

At the time, I was living a normal, carefree life, where I did not have to worry about this hidden price of fame and celebrity that haunts a lot of famous people. The only thing I had to worry about at this time in my life was getting up on time, going to my classes every morning, and going to my normal job that I held to pay my bills every month. Never in my life did I think that I might be heading down this very same path of creating a name for myself – until I formed Trimberger Enterprises in November of 2002, my freelance business – which started a media circus without me even knowing it.

Six years later and two years after I officially established the name of my company, I would find that I would be headed down this very dangerous road. As much as I wanted to get my story out to the world, I found that I had to be careful. Luckily, by a chance of fate, I had dropped the service on my landline phone and opted to use my cellular phone for everything two years earlier in the month of June 2002. I had read that a lot of people were getting rid of their landline phones and just using a cellular phone. I thought, what is the point of paying for long distance when I can get the same service free of charge with my cellular phone? Little did I know that by doing this, I was giving

---

[1] People who try and follow you and take your photograph for the tabloids.

myself a piece of sanity that I would not have had if I did not drop my landline phone – not to mention saving at least $100 every month.

This at least gave me a good line of defense against people who might otherwise try and look up my address and phone number by calling information. At this point, the only publicity that I had were the numerous calls that seemed to arrive at odd hours of the day from telemarketers, trying to sell something to me. Now, I don't have to worry about telemarketers calling me. Plus, it is easier on me professionally. I can just set the phone to silent mode so that the voice mail picks up. I won't even hear the phone ring. I found that by doing this and just using a cellular phone, I would not be listed in any directories, since cellular phone numbers don't appear in the white pages – they are not even included when you dial 411 to look somebody's name up. There are a lot of people who work within my profession who don't have a landline phone for this very same reason. They are trying to minimize the exposure they have to being hurt by angry people, and then some are trying to avoid the paparazzi.

With the numbers of people dropping their landline phones, cellular companies are now getting the business that the local phone companies used to get. Not as many people are listed, unless they choose to be listed. Certain government authorities are hard at work trying to change that rule and make cell phone numbers listed, but we will wait and see if that ever happens. It would be a phone book which has the listings for cell phone numbers – but you are not going to find a number listed under my name in that book. The only people who will have my phone number are the people who have a business reason for having my phone number. My family has already figured out that calling me is futile, especially when "campaign" season hit a feverish pitch after my literary agent had a search warrant served on her in January of 2004 and I was getting little sleep just so I could get my work sold a lot quicker.

I never knew that two years later, this might end up preventing another type of call. It would prevent, to a certain extent, people who would want to look me up to terrorize me or my family, scare me, or try to get some attention for themselves. With the advent of the internet, I know that there is no way that I can be found. I personally looked my own name up. I actually was thinking that perhaps people needed a way to contact me, but after going to the mall and being addressed by first name by complete strangers, I immediately yanked my name out of the directory.

I am positive that there are people who were angry when they tried to locate my number and were told that nobody with that name existed in their database. I exist. I live in the southwestern part of the country – I am not about to say where in the southwest that I live.   But directory assistance doesn't know that. And if you happen to know somebody, who knows somebody, who knows me, and happen to be able to access my phone number that way, that is also monitored. I assume that people who might call blocking their identities are not dumb enough to leave a message. You want to know who I am, but you don't want to reveal any information about yourself. Okay. Let's level the playing field so it is fair. If you want to play the cat and mouse game with me, I will attack back. You can call my number all you want – just don't be dumb enough to cross that line that separates the smart from the stupid. If you cross that line, you may want to take occasional glances behind your back.

A lot of people don't understand how much work it takes to get a screenplay sold, especially when you don't have any track record. I probably wrote well over a hundred letters, spent who knows how much money on postage, and had not one request for my script. Recently, I had changed my logline to say "An accountant becomes a teacher." I had a few requests to read the script from that. Believe it or not, a lot of the time, the decision to buy a film is made solely on the logline for the movie.

With no agent, no manager, nobody saying that my work is worth looking at, what I needed was a producer who saw the merit in my work. Unfortunately, a lot of the time, producers tend not to want to look at any work that does not come through an agent or a manager. There is also that catch-22 situation. No producer wants to look at unsolicited work, which doesn't come through a manager or agent, and no agent or manager wants to take new writers. My father was really happy at how difficult things were getting for me. He didn't like the idea of the fact that I was following my heart and doing what I liked instead of doing something that somebody else wanted me to do.

THE BOOK FESTIVAL

One weekend, my friend and former neighbor, Lynn Wiese, was going to be sitting in a booth at the Arizona Book Festival and she said that I might want to come and check it out. I get there early in the morning, and spend the entire day there. Since she and I live about 100 miles from one another, I don't see her very often, so I decide to go and spend an entire Saturday talking with her since I had nothing on my calendar at the time. She and I take a break as she shows me the different booths that are set up. I walk around all day and do not take off the dark glasses that I have brought with me, even when the sun eventually goes down. As I walk past people, I make direct eye contact, and I know who may know about me. I notice that in many cases people stare at me – giving more than the type of glance that would be accorded to a stranger. I have no way of describing what I felt that day, but I was experiencing a weird vibe that entire day that I was with my friend.

Handing my business card to someone who she knows, it hits home how many people know about me. As I am walking around looking at the different booths, I feel like I am being followed. I am feeling a certain energy source coming from somewhere, that makes me feel very vulnerable. There are a few people that walk by Lynn's booth and look me straight in the eyes – which is very unusual given the fact that not a lot of people were stopping in Lynn's booth that day.

That same day, Janet Napolitano was at the book fair – as was evidenced by the presence of the U.S. Secret Service. What I would give to have armed security guards looking after my interests. Since I am not a member of the government, I don't get that luxury. If I want to have that level of protection, I have to pay through the nose for it. Next time you wonder why the price of that concert ticket seems a little bit high, factor in the cost of personal security for the singers and the band. It is not a cheap cost by any stretch of the imagination.

There are some people that pay upwards of a quarter million dollars a year for what is called executive protection[1]. Until you get to the point where every Tom, Dick and Harry knows you, you will not understand the point of executive protection. At least Janet Napolitano does not have to pay for her armed security guards since she is an employee of the government. For those people who think that I am being a jerk for paying for people who have a license to carry firearms or know martial arts so they can push me out of the way of an oncoming vehicle that is attempting to run me over, then follow me from place to place when in public, spend a day in my shoes. You would be out of your mind as you start to worry about who may or may not be following you, talking about you, or pointing you out to people. It makes the hair on the back of my neck

---

[1] Protection that is given for celebrities and public figures.

stand up – well now the playing field is leveled so things are fair. Call it what you want. But don't make judgments about people who are living a different lifestyle from yourself. Until you are at that point in life, you have no right to be critical of those people. After all, it is people like myself that create jobs for those on the lower rungs of the ladder.

In a lot of cases, the people who want to get themselves close to me don't want to know me as a person. They only want to know me because I happen to have a lot of money. Well, guess where that money goes? It goes to the people that matter in my life – my managers, my attorneys, security companies, publicist, agents, and those who played a role in getting me to the point that I am at in my life, not to mention my family, who I plan on treating to an all expense paid trip through Europe when the right time comes and I am financially stable again like I was in Wisconsin. At least it will get me out of the line of fire for a period of time.

CELEBRITY STALKERS

There are some out there who think that they have me eating out of the palm of their hands. I understand that not everybody is like that. But the saying, "One bad apple spoils it for the bunch" holds true. For all of you people that think that you can play the cat and mouse game with me, I know for a fact, that you wouldn't have found out about me if it were not for my internet presence. Perhaps you did not know, that either my security company, or myself, can run reports on anybody who visits my website. I can see the keywords you used to find my site. I can tell from looking at a report, what words you typed into Yahoo, Alta Vista, MSN, Internet Explorer, Netscape, or whatever the search engine of your choosing happens to be.

I can see whether you typed the words "tegretol and osteoporosis" or "tegretol and bone density" or "trimberger" or "personal autobiography" or "Euphoria" or any way in which you access my website. Any search words that you use to find my site will be displayed in a query that I run on my computer. Based on this query, I will know whether you used search queries, in which case, I need to do closer scrutinizing, or whether you accessed my site by directly typing the URL into the internet browser. These days, very few people come across my website by accident because the word has gotten around about my story. In a lot of cases, my site has already been bookmarked or somebody typed the URL to my website into their browser. A lot of these reports, I print off every month and review them to see if there is a pattern change that I need to be concerned with.

I can tell if you typed the URL to my website, in which case, you either know who I am and I have given you my business card or you perhaps already knew that my website existed. I can tell what the first page of my site you accessed first, and where you went after that. I can also see the *region* of the country or the part of the world where my site was accessed from, and from *what country* you accessed my site from. (Yes, I have had several visits from Australia, The United Kingdom, Canada, and France, but I am not about to reveal who those people are. I have ways of finding out exactly who is accessing my site. If you have nothing to hide, this shouldn't concern you too much. After all, you want to know who I am and what my patterns are, right? It is only fair in return that I know who you are!) You can run from me, but you can't hide from me. I will find out who you are if I have the need to know that information.

For example, I know that I have seen a hit on my website from Senator Kerry's administration in the Washington D.C. area. I also at one time had my site visited by Leonardo DiCaprio himself, after his agent forwarded my letter and story on to him. I can piece together certain things to figure out all of this, based on internet domains, exactly who is visiting my website. It is a brave new world out there. Get used to the idea – that if I want to get the information about who is visiting my site – and I check my website statistics on a daily basis – I have a way to do that. I even knew when my attorney visited my site before he even admitted that fact to my attention when I met him in California.

I determined that I was in good hands with my production company – the one that passed on it because there was no market – when I ran the statistic report, and noticed that there were fifteen hits to my website on June 26, 2004, a few of those hits being from Canada, where a lot of Hollywood movies are filmed to cut down on budget costs. With that being higher than usual and only a month after I submitted my screenplay to them, I did some further checking and found some interesting information that I will not reveal in print.

This example should go to prove that if I want to discern who is accessing my site, I have ways to do that. All I do, is type in the domain name along with the identifying information and I can see *who* was on my site if I need to. If you want to keep your privacy, short of crawling under a rock and remaining there for all eternity, maybe take that big box that is sitting on top of your desk and throw it out your window, because that is the only way in which you will remain anonymous to those of us who are computer literate. After all, my life depends on it.

I can see the number of hits that I am getting in a day, in a week, in a month, or whatever I choose to see. Based on seeing the length of time that people spend on my site, I can tell if I am getting a majority of people who have stumbled onto my site by accident and quickly exit, which means that they only stay on for under five minutes. If they spend more than ten minutes on my site, I know that my story interests them and they stay on my site to learn more information about who I am. This is an important piece of information from a marketing standpoint, because the people who stay on the longest, usually are people such as producers, agents, and those with a genuine interest in the story that I propose. A lot of these people spend close to thirty minutes on my site.

I can run another set of reports that tells me what page you entered my site on, and where you exited my site. This is important since it tells me more of who is accessing my website. If you entered on the front page, chances are, you did not use a search query to find my website. If you entered anywhere else on my site, you probably did use a query and found my site that way. Within this same plethora of reports, I can see what pages are getting hit the most, and what pages are getting hit the least.

I will never forget my father's parting words as I departed from the airport in Milwaukee. He said to me, "go and get that job, Eliot." That spells one thing to me – boring. He may also be trying to divert my attention so I don't find out something that he has to face for his remaining years – that it very well could be possible that childhood vaccinations played a role in me getting epilepsy. What was never communicated to me, was that my mother was about to lose her job, which I would find out a month after I got back.

I imagine that I would get people saying that I have nothing better to do than sit in front of my computer and spy on other people. Believe me, I have better things to do than spend all my free time figuring out who is on my website. From a marketing standpoint, however, I need this information. If I did not have this information in my arsenal, I would not know what I was doing right and what I was doing wrong. For example, just recently, I have noticed two separate hits using my attorney's name to find my site. That means that somebody typed in my attorney's name and found my site. Noting the time that these searches were run, with the number of hits that I had shortly afterwards on my website, that could mean many things. Chances are, some producers were probably contacting him to get a copy of my movie script. Again, I could be wrong. But it is the first time in a while that two queries using the phrase "joey sayson attorney" from two different parts of the country were done. What followed a short time later were more hits than usual.

In the middle of August, 2004, I had followed up with the entertainment company, trying to maintain my patience level. Shortly after I got an email back in response to my request, I later on checked the stat reports on my website. I check the daily reports, and there were nineteen hits to my website in a single day – all blocked because whoever accessed it possibly had the cookies turned off. Whenever that happens and I don't get a visitor detail report on the screen despite the increase in hits, that tells me that there is a good possibility that the person who was visiting was form a studio or production company, and understandably, wanted to keep their identity confidential. With the internet comes a lot of security issues as it relates to information. There are reasons why producers and studios like to be secretive about what they are doing and I respect that. It has to do with preventing another studio from ripping off a project right from under their nose in the heat of a bidding war. Trust me, it is so complicated, that even with myself, I don't know everything with regard to why studios and production companies keep secrets like that. I am not yet in the environment. Until I get a promotion to creative consultant, I am just writing for the industry.

Since you don't tell me who you are when you visit my site, but you want to know more about who I am, I figure in all fairness, I should have a right to that information. For example, from looking at past reports, very few people went to the page on holistic medicine. Those numbers actually increased in August when all these hits came through on my website. That tells me that perhaps there is limited information about holistic medicine. A lot of the time, the people went to the pages that were directly about the book that I am writing, or to the page that has the progress for my movie. These reports tell me where I need to improve my website, as well as to give me a general sense of who is visiting my website based on location, time, and the number of return visits to my website.

So, if you think that you know so much about me and my patterns, that makes two of us – because I know exactly who you are. As the old expression "fight fire with fire" says, if you think that you can hide behind your computer and remain anonymous, tell that to somebody with technical skills that would surpass the highest security professional. I can even see how often you return to my site and how regularly you return to my site. So much for hiding behind that computer screen, thinking that I can't see who is looking up my website. When you realize that I have just as much information about you as you have on me, that might make you think twice before I have no choice but to raise the price of poker. The threats against those who are in the public eye is very real. The following example illustrates exactly my point.

IDENTITY THEFT

If someone asked you for your bank account numbers would you give them out? Would you give your social security number to someone who asked for it over the phone? I would assume that the answer would be "no". Why? You don't want someone assuming your identity and taking out a loan for a house or a car in your good, pristine name. That is understandable, because this very thing happened to a CEO of a Hollywood studio. According to a book titled *Your Evil Twin* , written by Bob Sullivan, he talks about the dangers that are faced by those who are vulnerable to having their identities stolen. Specifically, he gives an example of a CEO who received a package at his house one day. The person who contacted this individual had sent the producer in Hollywood a copy of his social security number, his bank account number, credit card number, and part of his credit report. Included in this were information on all sorts of people in Hollywood which contained personal information not meant for anyone else to see. It is understandable to an extent why I would be raising this issue. If somebody were to see a huge check with my name on it, it wouldn't take long for my address to start circulating to everyone in town.

Yet, in this country, we have something called the paparazzi, which seems to have no problem with invading the privacy of those who we consider to be a step above the rest of the population. In all fairness to celebrities, I have to say that this is also an invasion of privacy. On the other hand, those who put themselves in the public eye have something called the right of publicity which means that the only recourse that those with a publicity status have under the law is if they can prove malicious intent on behalf of the other party. A lot of times, these cases are thrown out of court because those who are in the public eye have a certain expectation of publicity and therefore, don't have the same expectation of privacy that an ordinary person would have. However, this does not mean that you can ignore the rules of civility that society has prescribed.

I am not an attorney, so do not take the above information to be construed as legal advice. I am just speaking from personal experience. I can say this, because I am able to have empathy with regard to the invasion of privacy of these individuals. I have never been followed by the paparazzi, but there are a good number of ordinary people who do the same thing the paparazzi does – they pick them out of a crowd. What you don't realize, is that when you are famous, and your name is out there on your own personal website, there is a certain part of your life that is taken from you. That is why all these famous people eventually employ bodyguards around the clock.

As a general rule, people are concerned with identity theft, about somebody stealing our social security numbers, about somebody misusing our good name and getting credit in our name, about being stalked by somebody who is hell bent on hurting us – but they don't even think twice about invading the life of somebody who we don't even know – all because celebrities have relatively deep pockets. I am not saying this about everyone reading this book. I assume that the majority of people are honest in their intentions. But, I have been given a little too much recognition recently in the wake of my movie going into development that I have to let people know that I am on to them. One bad apple tends to spoil it for the bunch.

Common sense should tell you what might happen if you use mental terrorism as a weapon of choice – or worse yet – commit extortion by taking me or somebody in my family hostage. If you do something like this, I will be sure that you spend a long, long time in the graybar hotel.

There is a general attitude among the public that goes something like this: "Well, so-and-so makes a lot of money, so that just goes with the territory." Wrong again. People are not compensated at a high rate of pay because they expect to be harassed by other members of society or the paparazzi. That is not why celebrities are paid so much. If all someone had to do was get everybody talking about them in order to make huge amounts of money, then we all would be rich. But then, of course, we would not need to have expensive schools that are primitive in nature and don't let people explore their creative endeavors, which is what eventually will lead me into something that I talk about in Chapter seven with regard to holistic medicine.

Everybody has a market value on their heads, myself included. If what they offer society can't be found somewhere else, they make a lot more than the average wage slave. I was a wage slave for ten years longer than I needed to be and I have every right to say this. I am sure that when my bank account has seven figures in it, my father is going to be taking back every nasty word that he said to me and kissing my ass, all in the name of trying to squeeze money out of me like an overripe piece of fruit. And I am sure that he won't be the only one who tries this little trick on me. I will be sure that I don't fall for it.

That being said, we all have a right to a private life, whether we are famous or not, whether we have the bank account of Bill Gates with extensive benefits or the bank account of wage slave junior. I have never been a celebrity stalker. I never will be a celebrity stalker. I look up to certain celebrities – we all do – that I respect. They are the ones that make it easier to escape from our own hectic world every weekend. Celebrities worry about people who are looking to commit extortion, kidnapping, and are usually guarded about people approaching them in an inappropriate way. Before I became well-known, I had no idea how difficult it was to maintain that line between a normal life and the life of a celebrity. When I crossed over to the other side, that was when the shit hit the fan – without even knowing what I had signed up for.

Keep in mind, that when I first started writing this book and establishing an internet presence, I never expected to become a celebrity from it. That just happened by surprise – and it blasted at the core of who I was as a person. Now with a lot more at stake financially, I know that there are people out there who may try and use extortion or blackmail to get money out of me. The only person who is getting a big chunk out of my bank account is my attorney and other's associated with my business – because they are the ones who are working hard to make sure that I get what I am entitled to by the Hollywood industry, as well as the ten percent of my paycheck that I fork over to my manager who is responsible for taking care of the other business aspects that I don't care to be dealing with on a daily basis. After all, when you reach a certain stage in life and don't realize how different you are, you need someone who is going to be behind you one hundred percent.

While you can go about your life without worrying about somebody profiling your every move, it is not the same on this side of the fence. When I see a celebrity on television that I admire, it inspires me – to be like them – making a name for myself. It doesn't inspire me to terrorize them. People that do that are on a power trip.

THE DEATH OF A DREAM

September 12, 2004 – I was sitting at my computer when I see an email come through from Go Girl Media, saying that they have read my script, but there is not a market for the script. As a result, I desperately looked for ways to revive this fire. With no job, my parents out of work, the studio out of the picture, and the level of frustration growing, I did something out of desperation. I just blasted my query to over two thousand producers and executives the following day, anybody who would read my script. For every day that went by, I was starting to go more and more insane. I wanted to be able to cut my parents off. Unfortunately, the same people I was growing to dislike, I needed to help finance a dream that very well might never happen. It is so irritating that at times, I just want to crawl up into a ball and die.

Everyone seems to have their reason for not cutting me an option check for ten percent of the value of what my screenplay would be worth. There is some truth to the saying that when you don't have any kind of track record, it is difficult to get people to listen to you. When I write my next book, things will probably be a completely different tune.

ILLUSIONS ARE DECEIVING

Don't mistake me as a mental case that belongs locked up in a mental ward because I seem to be going off my soapbox. I am thinking on the same level that somebody who would want to do some harm to me is thinking on. The people that do this generally are trying to get attention, and I always ignore them.

While I am on the subject, when you come across a celebrity, don't try and introduce your daughter to them, thinking that she will be happy to be with somebody who has deep pockets. That person who you are introducing your daughter to, is already on the defensive fending off people who might make their life a living nightmare. Aside from being on the road more than half the year giving interviews with regard to my movie, talking in bookstores and other places about my experience, and giving interviews on talk shows, I am already going to be worn out to the point where I don't even have the time for myself that I need to have. That doesn't make for a pleasant homelife – at least right now. And mark my words – when my screenplay is optioned or sold, settling down is not going to be my priority, especially given that I will be in meetings just about every day with producers with regard to rewrites for my project – not to mention the daily conversations that I will be having with my manager. The fact that some people out there have done this shows that they have *no* idea about the dynamics of living life as a celebrity.

While I am on the subject of having a manager, so you understand, just to get my screenplay into a studio and into pre-production, I will be going over extensive notes from other people with my manager. Everybody involved on my project is going to have a say into how the screenplay should read, even though they did not write it. This is especially true for bankable characters, any producers involved, and the agents that represent the actors. I will probably be going out of my mind after my script sells, mark my words on that. It is a brave new world, even for me. And from what I have read, the amount of time that it take between signing the papers and getting paid can be many months. It takes forever to do some of these deals.

I have enough work-related problems of my own and the last thing that I need to be thinking about is who might be looking for a way to win the lottery. Maybe you should play the lottery if you want to win the lottery. Last I heard, the odds are about one in one hundred million of actually having the winning numbers. This whole experience will probably result in me moving to safer quarters. The unique nature of my circumstances doesn't make for a safe environment right now, which is why I probably won't be in Arizona much longer than I need to be.

BACKSTABBERS

Put yourself in my shoes, or in the shoes of any celebrity for a few weeks. When you have to abruptly shift the times in which you go to the store to get groceries, doing most of your tasks in the early morning hours – all at the cost of being real well-known because of your original story – you may take the same precautions. There were times that I would stay in my apartment and not ever emerge except if there was a national crisis that needed to be taken care of, such as getting Cozy, my personal guard cat, more food just to keep her from waking me up out of my slumber. When that cat wants something, she has a way of getting attention focused on what she needs. When I would be asleep for hours at a time because my eyes were on fire, she would meow to the point of hearing loss.

When she puts her fish hooks through my skin after I try to no avail to pull out the mats in her fur, all while standing on her hind quarters putting her front paws around my hand, holding on for dear life like that of a bulldog, it had on more than one occasion resulted in a trip to the refrigerator to apply some more aloe to my hands as a result of the scratching. This cat is so big that I should have probably picked up a dog from the humane society. The beast weighs about twenty-five pounds. Maybe a little less than that, depending on what kind of mood it happens to be in on any given day.

I can't count the number of times that I am in a store and I hear words that I know pertain to me. I am not going to list those words because I don't want to give anybody any satisfaction that they may have been able to scare me. When you are walking out of a grocery store and you hear someone say things about you behind your back, which was in the wake of the January 2004 incident when my literary agent was served papers for dishonesty, you start to worry about profiling. You start wondering if somebody has it in for you. If you have something to say about me, or to me, have the dignity to say it to my face like a man. Only cowards say something when a person's back is turned. This is something that has happened on more than one occasion. Usually that is because they don't want to be identified. The same thing goes for that one fingered salute that has been pointed my direction a lot more lately. Whoever is doing this, is somebody who has no dignity.

I have made the habit of looking everybody straight in the face when I think I am the subject of their gaze. My father calls that rude. I call that a self-defense mechanism. The same person who practices martial arts is the same person who was raised with the ideals that looking at somebody straight in the face was considered disrespectful. Back then, children were seen and not heard. Speak only when spoken to. Times have changed, my friend. Welcome to the brave, new world. We'll see who is right if and when this media circus finally comes to a close. Once I get the power to get the pandemonium going, I will make sure that my name never dies until I do.

Perhaps you are wondering why I have diverted down this path in my book. I do this to drive a point home. After a while, you would start to snap because of all the craziness, which is involved with everybody knowing about you in one form or another. You no longer have the personal space that you used to have in the months before you decided to let the world in on your story. Some may argue that I made my bed, so I should sleep in it. While this is true, there is a limit to what is considered to be acceptable by the general public with regard to the respect that should be shown to those who are on a higher rung of the social ladder than most people. I don't mean this as any form of disrespect to those who are ordinary people. At one time, for an extremely long time, I was there, so I know exactly what that is like to be an ordinary person. However, there is one primary difference that gives me this elevated status – I worked my tail off for it, and I should be able to reap the rewards without being judged unfairly by other members of society.

There was one time when I was home in Wisconsin, and was at a movie theatre – and I was waiting outside while my brother used the restroom. I was getting a bad vibe from somebody who was looking at me. I could read the look in this person's eyes. He was just looking at me for a little bit too long of a period of time. As my back turned – and again, he was out of my line of sight, lucky for him – I knew that he was going to judge me. My convictions were correct. He muttered a comment that was meant to show that he was not of approval to the work that I was doing and wanted me to know that he disapproved. Like I said in the beginning of the chapter, we all secretly judge one another for one reason or another, and we all have our faults. I have already revealed my faults in the previous chapter.

It will only come to a point when the problem gets worse. What they may not have known, is that I had people watching over me, who were within close proximity to make sure that things didn't get out of hand. If blows had to be exchanged, I wanted to be out of the situation entirely and let somebody else clean up the mess. *(Disclaimer: I am not a violent person by nature. If I am not given a choice in the matter, however, others will take any steps necessary to defend me from getting hurt.)*

I would work day jobs doing anything that was available – just to make extra money. It was so bad that I took any job that was available. As I walk into the building, there are people staring at me. Although I have worked at this company on a temporary basis one year earlier, there was a vibe in the air that I was picking up. During my lunch break, when I was reading 135 pages of "notes", someone comes up to the table and asks if my name is Eliot. I had worked there a year earlier. But, I had never seen this person before in my life – and this one specific place uses a lot of temporary employees when they need help with something. The only way that they would know me is through word of mouth or through my website. I look at them, and say yes. Glancing at them squarely in the eyes, waiting them out to see if they would tell me who they were, they backed down and walked away, because they don't want to expose who they are. I had a mental note already made of what they look like since I looked them in the eyes. A lot of people are afraid to be noticed for one reason or another. It seems that people want to know who I am, but they don't want to disclose any information about themselves.

The people who I was reporting to were cognizant of the fact that I was somewhat high profile. Actually, when I am doing my job, I usually am focused on that and don't make it easy for people to bother me. It seems that a lot of people don't want you to know who they are. Those people are cowards. This is mild compared to some of the more notable situations that have arisen as they concern celebrities.

I will give you an example of the type of craziness that exists out there in a celebrity's life where it can turn ugly. I was reading Michael J. Fox's memoir *Lucky Man* and he talked about how his wife was getting numerous threatening letters during the time that he was shooting a movie overseas in southeast Asia. Fox's wife was getting numerous threatening letters sent to her over a two month period, threatening her life – all because this person wanted to be involved with Fox.. His wife was told that if she didn't divorce Fox, that she was going to be murdered. Where there is a lot of money, there are gold diggers. And don't think for a minute that I am not on to all of these games. As a result of these letters, according to Fox, Gavin DeBecker, his security company, kept close watch on the situation and interviewed this person to find out what they wanted. I do take this stuff seriously. If it weren't for the fact that my lifestyle had changed significantly, I would not have given the story so much thought.

I was watching a show on Court TV called *Celebrity Stalkings* and they were talking about how the people stalking celebrities get the power from inflicting what is called mental terrorism. That is why Steven Spielberg, a director of numerous movies, has a lot of security around his mansion – because he has been put in the public eye – and has been the subject of a stalking. In the days when stalking was not against the law, this happened on a regular basis to celebrities. It was these cases that resulted in state laws that were passed, at least in the state of California, that made stalking someone a criminal offense.

The anti-stalking laws that were passed in this country were passed because of what celebrities have endured with people following them and threatening harm to them in some way. So, before you complain about how much money myself or anybody else who is a celebrity or well-known person makes, just keep one thing in mind. Celebrities were the ones responsible for the anti-stalking laws that were passed, if not in all states of the country, at least in the state of California after a well-known figure was murdered.

A fan who was obsessed with this person located the address to this person using Department of Motor Vehicle records and put an end to their life. Around the time that this happened, all a person had to do in order to locate somebody was to go to the local DMV office and ask for their address and pay the fee and they would get it. Whether or not that has changed since then is unknown.

A SCARE

In December 2003, I had somebody bang on my apartment door for twenty minutes straight. I looked outside, and there were two people out there, who I did not know. I called the 24 hour security in my apartment, and they told me "we don't respond to people knocking on doors." That is a ridiculous policy. Did it ever occur to you that a lot of home burglaries are committed when the intended target is home? Do you know why that is? If the owner is not home, it takes longer to find out where the valuables are kept.

I didn't recognize them – I called the police and filed a report. The same person came and knocked on my door at 6:00 in the morning. What he did not know, is that I happened to be up at the time working on making a huge sale so that I could get out of the dump that I am living in – and get to more secure quarters. Thank God I was smart enough to get an apartment on the second floor. Especially with the number of burglaries being up, a thief generally wants to get in and out quickly. They don't want to deal with climbing a flight of stairs. It only makes it more difficult to carry things out if they have to take longer to do it.

I got a physical description of him – and put a report in to the police computer system about him. Since that time, I have not had any other problems with that. But, do people like me a favor. If you are going to have an after hours security at the apartment complex, don't tell those who call that you are not going to answer certain calls. If I would have been injured, you could bet that I would have sued.

WHY DO I DO THIS?

Why do I devote an entire chapter to going off my soapbox? There is a moral to the story. I have gone through such a transformation, that I am unable to relate on the same level to a lot of people that I used to talk to, not to mention unable to relate a lot to society. Although I am still friends with those people, they would be unable to understand the other person that I have changed into. I want you to understand what it is like for somebody like myself to go about my day to day activities with people talking smack about me, trying to bring me down, especially when I don't see them putting their neck out there trying to get change started. The people who make these comments are usually making these comments out of ignorance. If they only got to know who I am, they would be able to understand the method to my madness. There are people who say that I should be the one to initiate contact.

Until you are at that point in your life, where everybody and their pet poodle knows who you are, you have nothing to complain about. Also, I had changed so much in the past year, that I was unable to even relate to some of my friends who I met after moving here. Hell – I barely even knew who I was anymore. What in the world is happening to me? A lot of the people who I had met only two years earlier, for the most part, didn't know who I was any longer. It was hard for me to relate to them. When you have learned the industry jargon, done research at the library, and are spending every day writing or working on different projects, other people outside of this circle take a back seat, and they don't understand that. My mother would be upset about the fact that I wouldn't call on the weekend. Why would she want to talk to me? She wouldn't even understand the language that I talk in anyways.

It is hard to explain to somebody what it is like to all of a sudden have to deal with the price tag associated with being a celebrity – when you don't even have the money involved with playing the part. Everybody wants to become a celebrity, be popular, be in the spotlight – myself included. When that happens, there is fallout to be dealt with in some form, usually in the economical sense. Then there are those people in my immediate family who are telling me that I should just "get a job". That is understandable – it keeps the cavalry away from my doorstep.

My mother, when I was home one summer, resorts to her usual ways, which are smothering me to death when I am used to coping on my own, taking me out on a walk, which is code for "I want to say talk about something that I know that you don't want to talk about." She would drag me over to the lakefront of Lake Michigan and see if I got some inner meaning from it. I have been living in a different region of the country and get my meaning from the mountains and from the sunlight. To me, that was like trying to deny the fact that I have changed. By showing me the lake, and saying that it reached out to me, I was sensing some sort of inner message that she was trying to get across to me.

My mother sometimes has this very rude habit of calling me and leaving a message on my voice mail telling me what everybody is doing. She does this and wonders why I refuse to answer my phone when her number comes up on the screen. In my own opinion, she is refusing to accept reality by calling to tell me what everyone else is doing instead of calling me to show an interest in what I am doing. It has gotten to a point where I have to be very picky about the friends that I make. Despite the fact that me and my parents have their differences, for the first time that I have seen, when I was home during my ten year high school reunion, it was the first time that my dedication to my work made my mother break down. I guess I must be doing something right.

There are people who actually show support by asking me what I have been working on. How do I know if somebody is truly interested in me? When they ask about me instead of talking about themselves. I have even been guilty of this on occasion – and have had to change my behavior to fit the circumstances. When I went back for my ten year high school reunion, there were people who had been places that I had not been and was curious to hear about what they had going for them and where they were working. They were people who had experiences outside of my own. They had been to other countries, which is something that I have not yet done. There were people who had been going to school in Europe, Asia, or some other area of the globe and had insight that I could not draw from because I had yet to experience that on some level.

PERSEVERANCE PAYS OFF

Then, of course, there are the things that some people simply are not able to understand with regard to what it takes to be a success overnight, if not yet intellectually in the form of a produced screen credit, but in the form of being well-known through an internet presence that has experienced a good number of hits the first year it was up. The countless hours that you spent up at night trying to save your business from taking a nosedive to the ground, trying to get somebody to buy your screenplay, the twelve hour trip going between Arizona and California, six hours each way, to meet with a working producer nice enough to sit down with you for two hours to help you to figure out what the hell you are not doing right and to also meet my entertainment attorney, who is somebody who will be there for me when clear sailing comes and I have a movie contract landed by a production company.

There is no way to get the average person to understand the level of commitment involved when it comes to making the correct decisions about hiring the right people to do work for you while knowing what decisions would be disastrous for the long term financial health of your company.

People don't understand something else as well. I sacrificed a lot of sleep in order to get my screenplay polished and marketable to the point where it could be sent to producers. I admit that depriving my body of necessary sleep was not the best idea – but I was hell bent on getting some sort of change implemented with regard to increasing the exposure that my freelance business had. Even as I am putting all this information together, my eyes are burning and I am lying down as I am typing all of this. Eventually, when the time is right, I am going to move to Los Angeles. I know that I have a fairly tough battle ahead of me. When I took the plunge into self-employment, following after my friend who wrote *Holistic Parenting,* I never realized that I was going to change on such a level that my own family would despise me in one form or another. Keep in mind that nobody else in my family has done anything remotely close to the projects that I am trying to sell currently. Without that family connection, I was an island in the Pacific Ocean.

My experience with my ex-literary agent taught me what to look out for and who to associate with in a way that no other lesson would be able to teach me. It also told me who my friends were. When I had put a note on my website with regard to my book, it didn't take too long for someone to come along and help me out. It ended up that at my ten year high school reunion, I met somebody with connections to an editor that worked at a publishing house in the twin cities – which may eventually provide me an advance so I can continue my work. If they did not care about me one way or another, they wouldn't have lent me a helping hand so that my story could be published by a trade publisher. The people who see that you are working hard and lend a helping hand are people who I know are good for my long-term survival.

When you are first starting out doing something that you have never done before, you are bound to make mistakes. A lot of the times, if you implement everything correctly, those mistakes are not made a second time. That is what delayed the publication of this book by close to a year and a half. That is what made me decide to self-publish the book until I could get an agent or publisher interested in the licensing of the copyright. If you take anyone who comes along who wants to represent your work, you end up learning the hard way that this world is full of people who like to take advantage of those people who have no idea what they are doing. You only get to make that mistake once. If you make that mistake a second time, then you really don't have a clue about what you are doing. This world is full of sharks and people that like to make life miserable for everybody.

These are the things that a lot of people may be unable to understand about why a person has a lot of money, why they are compensated at such a high rate of pay compared to the average person who works a nine to five job everyday. I don't know at what amount the president is compensated. If they are doing their job, running the country effectively, they should be getting everything they are entitled to. Because I don't know how to keep this country out of a war with foreign countries, I am not paid at the same rate because of the knowledge that I lack in that area. Unfortunately, we have an administration that likes to start wars and get other people mad. Where is John Kerry when we need him desperately? Hopefully by the time this book is published, he will be sworn in as President.

I also do this to explore what goes through the minds of the average person who, like everyone else, wants to be part of that invisible spotlight, which follows a celebrity around everyday. I will be the first to admit that when I was growing up, I wanted to be a part of that spotlight and get recognition. I was battling a problem that nobody knew anything about, with regard to the cause, or for the most part, how to treat it. I didn't even know what I was dealing with, or how to educate people in regard to seizures. I had to rely on other people's accounts of what happened.

With a first time writer, that normally would not make one a celebrity – except for the fact that I had a story that Hollywood wanted to hear. Even though I have gained a lot of knowledge, I have had to explore how to go about dealing with the public knowing me on a closer scale than I had imagined possible. It has ripped the core out of who I happen to be on a personal scale. If nobody understands me, I have no problem being alone. However, I know that there are honest people out there somewhere.

THE RULES

For anybody out there with the idea that they are going to find a way to marry my money, or sue me for my money if that is the case, I don't plan on making life easy for you. If you had an enormous amount of assets to protect, I think that you would probably do the same thing. Second, at this stage in my life, by nature, I am not as trusting as I used to be. And I am apt to look at any "invitation" to play house with an increased level of suspicion. I have a lot more to lose financially. Cozy and I are just fine and we like it the way it is without any more drama entering the picture. If I decide to change my mind on this, I will let you know.

You can go about your day to day life with a certain measure and degree of privacy. I can't do that, because of my internet presence. But before you tell me to dismantle my internet site, I will tell you that if I did that, I would not be where I am today. As you are reading this, chances are, I have retreated somewhere in Europe to get back in touch with my roots.. A very extended hiatus is going to be in order very soon, because I need it so that I can recharge my batteries. I also owe it to my family, who sacrificed a lot during the time that I did not have a job. That is where my priorities lie.

Also, if you come up to me and ask me what my name is, at least have the human decency to tell me who you are by introducing yourself. I like to know who I am talking to. I am not another computer out there. Under the surface, I have human qualities just like the rest of us do. That will put everything on equal footing. You want to know if the person in front of you is the person who you think they are. Don't you think that I would like to know who it is that I am talking to? If you are not going to follow this rule, then just leave me alone. I have much better things to do than waste my time talking with someone who wants to remain anonymous when I am putting my neck on the line. As a matter of fact, that is the equivalent of insulting me.

Because of the circus that I created out of my story, I had to do something for my family, particularly my mother and father, who made huge financial sacrifices to keep me from filing chapter seven bankruptcy – which I had been threatening on many occasions to do if things didn't change on my end. With the financial security taking that out of the equation, I didn't have to worry about filing bankruptcy – but had a problem that was on the opposite extreme. It was one that would make me vulnerable to people who had it in for me. You can imagine the things that I have to worry about from a security standpoint. By the way, anybody who wants to get to know me has to be personally approved by Cozy. But don't worry – she won't bite until she is physically attacked.

However, shifting gears to another topic, the idea that the very people who were treating my case were against me seeing a holistic doctor gave me another perspective to talk from. It basically poses a question. Is one system of medicine better than another? Or, is it possible for American and holistic medicine to co-exist in the same playing field? I don't know for sure how to answer that question – because, at least for the time being, American medicine and alternative medicine cannot co-exist in the same playing field as long as the powers that be are making the rules.

Chapter 7

Doctors and Quacks

Driving 200 miles to Tucson, Arizona to see my holistic doctor, I finally am going to get the answer to my question that I have been raising with both my parents and my past doctors for over twenty years. It would be this trip that would also tell me something else. It would tell me that the path I was taking would be a lonely path, as I stepped outside the normal channel of medicine used to treat disease. It would be a decision that would put me into a position where I would have to fight a war on two fronts.

I was not sure at this point whether I should make my own situation better, or whether I should keep things the way they were. I have an internist who is compassionate and caring. When he first saw me in rehabilitation, it was clear just from the energy I was picking up off him, that he was a caring person. And I still think that of him, regardless of what holistic medicine does or does not do for me – I just dislike the way illness is approached under the current system.

And my health insurance, which paid over $50,000 for my surgery, cancelled me since I was an out-of-state resident. I had to find a drastic solution to my epidemic. Turning to holistic medicine to save my pocketbook from too much trouble way down the road, since I would eventually be paying out of pocket $800 a month for my medications, I wanted a solution to what has been a long-term struggle for me.

This trip would redefine my relationship with my parents, as I brought up the fact that my holistic doctor claimed that vaccinations were the reason that I have epilepsy. When they saw my appetite improve from these remedies, they could understand how it had a place in my life. It was after my internist made a recommendation based on my weight, that I see a nutritionist, that I set up this appointment.

For the first time in my life, I started to take the ingredient label on that ketchup container very seriously...

AN ANSWER TO A 20 YEAR OLD QUESTION

I think that I am going to get killed when I mention this. Here it goes. Every neurological disorder in this country, unless there is evidence of a head injury or something else to that degree, was probably caused by childhood vaccinations. Back around the time that I was born, a lot of the vaccinations had mercury, which is a preservative. Why do I say this? Because in the wake of my insurance being cancelled and being completely uninsurable, I had to come up with a reason for this madness that caused me to go into seizures without regular medication. I thought to myself, "If doctors are not able to say what has caused the problem, then there must be something that doctors don't want the public to know." And that is why I made it clear to The John Kerry Administration, through a letter that I sent them, that if he is elected, I want to see The American Medical Administration overturned or refined. They are no better or worse than the tobacco companies, by giving vaccinations to children for diseases that no longer affect the human population. Why are we continuing to vaccinate children against diseases that have been eradicated a long time ago?

A STINGING SENSATION

I remember when I was a child, there was a bee nest that was underneath the flower bed in the front yard of my parents house. Periodically, I would get stung by bees and would run into the house crying from the pain. I would have a welt on my skin from the bee sting. Twenty years later, when I am living in Arizona, I was watching a program on the Discovery Channel, which was profiling a case of somebody that was stung repeatedly by hornets over and over. Too many stings can cause death due to the venom. Ironically, that is something that holistic medicine does. By tricking the body into believing that it is sick, it produces a natural immunity to certain problems. The same poison that can kill a person is the antidote for certain medical problems. The following point is an illustration of this example.

Another byproduct of the stinging sensation that was reported after the victim was released from the hospital. While the victim was recovering in the hospital, she got home and noticed something different. The debilitating arthritis that she suffered from had gotten much better as a result of getting stung multiple times. What is actually poisonous can be the same thing that gives a certain amount of immunity to disease. That is one of the odd paradox's of holistic medicine. Instead of suppressing the symptoms that cause the disease, the poison actually creates a certain level of exposure, thereby creating a natural immunization to whatever causes the disease. What the poison does is trick the body into thinking that it is immune from the disease naturally instead of artificially.

Not too long after I got nailed with osteoporosis at a young age, I started to get a better idea of the relationship that exists between childhood vaccination and chronic disease. If it were not for the fact that I moved to Arizona, fractured my hip, and eventually met Lynn's holistic doctor, I would not have as much of an understanding as I have of holistic medicine. During this time, however, I was still under the care of my internist for the prescription of my medication to prevent fractures. As I was talking with my holistic doctor, it became clear that I was not even understanding his "language". In other words, the way he treated the problems that my body was going through was by treating the systems one at a time. First, he gave me remedies to help my digestive system. After that, I got another set of remedies that helped to clear my body of any cancer that I may develop in the future. As a result, I am screened for things such as prostate cancer for life. Only a few weeks later, I notice that my appetite has gotten much improved as I can stomach milk without feeling sick. Usually, the only thing I would drink would be Coke or Pepsi. Although I still drink those beverages, I also drink a good supply of milk fortified with vitamin D. It was actually my internist's recommendation, based on the fact that my weight was real low, that I see a nutritionist. At the time, I didn't know that this would be a holistic doctor – which may hold views that my internist does not agree with.

. Acting like an idiot, I walk into an appointment with my internist, which was a routine appointment, and I tell him that I know how I got epilepsy. Childhood vaccinations. The first word out of his mouth was "Quack!". He said something that is absolutely correct and I have to give him credit for that. He said that holistic doctors were blaming vaccinations on everything from autism to epilepsy to every conceivable chronic problem associated with the central nervous system. Why is it that the American Medical Association does not embrace holistic medicine from the standpoint of a person's general health? I will give you an illustrated example of why perhaps this is not seen by the American Medical Association as a cause for a lot of neurological disorders.

I read an article that recommended that women who are pregnant should avoid or reduce the consumption of tuna. Tuna has been linked to a lot of neurological disorders, which are linked to the mercury that these fish carry, a byproduct of water pollution. These big fish eat the smaller fish, which also have mercury, and as a result, the amount of mercury contained in the fish is enough to cause harm to a fetus. For a developing baby, the amount of mercury is enough to cause a long range of neurological problems.

That is exactly what we have when we vaccinate babies in their first year of life. The mercury, which is essentially a preservative that makes the vaccine last for many years, along with other chemicals, attacks the central nervous system and is the root of the cause of neurological problems. Around the time that I was born, mercury was used in vaccines on a regular basis. Recently, the Centers for Disease Control removed mercury from the vaccines that are used out of public concern. However, there are still other chemicals in these vaccines that cause long term problems.

The American Medical Association is very similar to the tobacco industry. By vaccinating every baby, essentially, they are guaranteed to have a customer for life with this approach. That is why you probably have never hear about holistic medicine. How can the medical industry make money this way? The American Medical Association does not allow licensed doctors to practice holistic medicine.

If I have my way, there will be a lot of people in the medical industry without jobs in the upcoming years. Because my plans are to overthrow the American Medical Association and get holistic medicine established. A war was started – and I intend on finishing it. You can count on that. I highly suggest that every sane person reading this book write a letter to their state representatives, telling them that you object to the use of vaccinations in your child and the pressure to vaccinate when it is completely unnecessary. We all came over here from Europe or Asia, where holistic medicine is routinely practiced. We all have roots in Europe, South America, or Asia. In Europe, where my holistic doctor is from, holistic medicine is universally practiced, as it is more friendly to the body.

In the United States, what do we do? We put money before people. We are the only country on earth that lets capitalism get real ugly. We, without question, allow those in the medical profession to decide what is in the best interests for our children. I think we can blame a lot of the nutcases in society on the medical profession. They don't seem to give a rat's ass about a person's welfare – all they want to do is make a quick buck. When was the last time you heard of a person dying from the chicken pox? Capitalism has indeed come to a dark hour. That is, in my opinion, how The American Medical Association got started – by rugged individualism. While this is all fine and good, there should be a certain obligation to the public. That is what led me to investigate the positive aspects of holistic medicine. It started me down that dark path to Tucson, Arizona, which would get me completely misunderstood.

Tucson, Arizona – December 12, 2003 – Driving 200 miles south from my sanctuary to see my friend's holistic doctor, I find out all about what holistic medicine is about. The first thing he does is asks me to stick out my right index finger. Rather, he was going to prick my finger and draw a sample of blood, which he would later use to put together a remedy that I would take twice a day. This was a medication that worked with my DNA and instead of suppressing symptoms, it would eventually get to the root of what caused the problem. In addition, he gives me this list of food that I am supposed to eat on a daily basis while I am taking these remedies.

The first thing I am thinking, is that these foods must taste really bad. They don't taste bad at all, as I would find out. The foods that I buy at the natural food store taste great. The reason – they are not loaded down with preservatives – which is what is put in food to keep them fresh and give them a long shelf-life. What you are really doing, when you consume the foods that are loaded with preservatives, is similar to what vaccines do. I will give you a good example.

When I was working at Aurora Healthcare in Milwaukee for two years, I would stop and get a sub sandwich either before I went to work or on my dinner break. I ordered a foot long sub sandwich with mayonnaise and mustard on it – and ate the entire sandwich voraciously. I got behind the wheel of my car, and went to my job. I felt fine as I got to work and settled behind my desk. Then the completely unexpected happened. Out of nowhere, I felt the butterfly effect again, so I quickly fumble around for an orange soda to drink quickly, hoping that the aura will disappear.

Before I got my driver's license, I found that drinking orange soda would help me when I had an aura, or pre-seizure feeling in my stomach. My seizures originate in my stomach. I had read about how people find using something to counteract an aura helps them. That is what I personally found with orange soda. In this case, however, it didn't disappear – and my body went into a seizure. The preservatives in the mayo and mustard were as strong as the vaccines used in the first year of a child's life.

My co-workers come over and ask if I am okay, and I say that I am fine. I don't have a migraine headache, and I am sure, that had I not stopped at Subway on my way to work, that I would never have had that seizure. I chalk it up to nothing. That is where the communication starts to get cloudy. I can tell my doctor that I had a seizure – but if I say what I think the seizure is from, it won't be confirmed. I am at a loss for what to do. There is no way to "scientifically" prove it, because medication has a different reaction on two people with the same condition. Doctors prescribe the same medication for anybody with certain symptoms, when they should really be addressing the cause of the seizures. Holistic medicine is what addresses the *cause* of medical problems, not the *symptoms* of medical problems. Symptoms are secondary to the actual cause. I don't have to have any type of medical training to tell you that. I will use my own situation as an example.

When I had experienced a fracture at a young age, the fracture was secondary. I had lived for fifteen years without my body showing any symptoms, because osteoporosis is considered to be a silent killer. You can have osteoporosis and not show symptoms until you end up in the hospital with a fracture. Science teaches us that there can't be a cause without an effect. In my own case, the *cause* of my osteoporosis was the prescription medication that I had to take for numerous years. The effect, was a low bone density reading that I was unaware of. Holistic medicine is a medicine that treats the whole individual without the need to see a copious number of specialists.

My holistic doctor is originally from the country of Denmark. Over there, as well as the entire continent of Europe, holistic medicine is universal medicine and is solely practiced, as it is seen to be friendlier to the body than American medicine. They also have national health coverage where everyone is covered by health insurance and generally look out for people. When I was in the hospital, I was referred to an osteopath, an internist, an endocrinologist, a urologist, and a nutritionist. Why not cut out the middle man and go right to the source? We have the most inefficient form of business in the United States. On the other hand, if it were not for this, I would not have the job security that I currently have today. Perhaps holistic medicine could be incorporated into helping those who would benefit from it. There are already some doctors in America that see the value in holistic medicine.

On more than a few cases, I have had seizures directly as a result of something that I consumed just a few hours earlier. The synthetic additives that are used in a lot of foods these days make the shopping mall a complete minefield. There was once a report on a television news station about high fructose corn syrup and the malady of effects that it had on certain individuals. This synthetic ingredient is similar to the aspartame that is used in Diet soda.

I am convinced that it was responsible for a seizure that happened shortly after pouring some of the ketchup on my plate to dip the fries that I had purchased from

McDonalds into. I look inside the bag, and they forgot to include the ketchup which they usually put in with my order. I grabbed some from the refrigerator and learned a little lesson. Since I had not had a seizure since the time that my car landed in the ditch after moving to Arizona, I knew this had to be the case. I was thinking, what could have caused my seizure? I had been getting regular sleep, even staying up all night would not be enough to cause a seizure to happen. I had been getting sleep at odd hours for the past year and a half. If a seizure were due to lack of sleep, I would have noticed the effect almost instantly.

Then it hit me like a ton of bricks. I consumed some ketchup with my quarter pound hamburger and fries – ketchup that had been sitting in my refrigerator for a long time and did not taste too good. It was watered down and generally had that bad taste to it. I was basically consuming artificial food. That discovery prompted me to read the ingredient label on the ketchup. The number two ingredient in this ketchup was high fructose corn syrup. I took this ketchup and threw it in the trash, where it belongs. Thank God I was not driving around this time.

Doing research on high fructose corn syrup, it was an additive put in there to prevent obesity. However, the food additive also seems to be responsible for causing nausea, seizures, and headaches from the study that I saw on television. This additive is in more foods than you may realize. The food and drug administration doesn't seem to be doing too much about this, either. Part of this has to do with the fact that they don't want to believe that there is any scientific basis for these problems.

Relating this to the chemical aspartame, which is put in a lot of diet sodas as a sugar substitute, similar symptoms show up after a person ingests it who is susceptible to seizures. In February 2002, a month before my fracture, my family visited from out of state and went hiking. Before we left the hotel room, I was thirsty. I had made the stupid mistake of consuming my mom's Diet Dr. Pepper. Everything that my mother consumes is diet, because she believes it will make her lose weight, something that is also untrue.

Everything was fine during the hike – until I got back down to the car. Jumping down the hill and getting back to the car, after this workout, I immediately have a myriad of symptoms as a result of consuming aspartame. At this point, I am turning pale and nauseated with a migraine headache. I was in so much pain, that I just wished somebody would shoot me. The thirty mile ride home was absolute hell. I can feel my body temperature increasing rapidly. It didn't help at all when my mother threw a jacket over my face. My face was burning up! I did not need some idiot that did not understand what I was dealing with to throw a jacket over my face which had already been started on fire. I think that ride home was the longest thirty miles that I can remember Thanks, mom – you just made my problems worse. You are lucky I only had a headache. That is something to think about if you are one of the millions of people who consume diet sodas. You have to take into consideration that your choices inevitably are going to affect other people who might inadvertently drink the soda and get sick. Then, I get back to my apartment, and take two Ibuprofen tablets, which push those symptoms deeper within the body.

SOME PEOPLE JUST DON'T LEARN

"Mom, just for the record, for my own benefit, would you throw out the Diet soda? I realize that you are on a "diet", but diet soda is full of synthetic shit that is doing unknown damage to the people that are important to you, and you may be unaware of that until it is too late. It does not do me any good. And you are not going to lose weight by drinking it. Maybe if you would find some way to chill out and quit bothering those around you, some of that weight would come off. Thank you for listening. I apologize for having to be rude."

What aspartame does is cause the neurons in the brain to get over excited, essentially killing them. One time in December 2003, me and my family went to see The Milwaukee Bucks play against the Pacers. I was not showing any signs of trouble – except from my mom's account. Traditionally, when we go to a basketball game, I always ask for a soda because I naturally get thirsty from all the cheering. I told my father that I was thirsty, and he took the liberty of getting a *diet* soda, which I was adamant about not drinking. He told me that this was better for me. What school did he graduate from? "Did I just say that I *don't* prefer diet soda, damn it?"

I got the look from my mother, as I brought this up to her later on. This was the look that said, "Shut up and quit making us look bad." Since my father was not going to get another soda, I just sipped it slowly. Around half-time was when my mom noticed that I was not my usual self. Becoming quite reserved, it may have affected me on some level. This is another good example of some of the idiots that I have to call my family –

and why I don't prefer to talk to them unless I absolutely have no other choice. Their habits are dangerous for my health. Honestly, I think they know this and they intentionally do this on some level to do harm to me.

Aspartame is responsible for over exciting the nervous system, basically killing brain cells. Aspartame is marketed under many different names, including monosodium glusomate and Equal. It has been found to cause nausea, seizures, upset stomach, migraines, fever, and other symptoms. There are those that don't agree with me. They are the same people that think diet soda will make them lose weight. The only way for them to lose weight is to chill out and not worry so much. This is not the only additive out there that causes these types of problems. On the news, high fructose corn syrup was talked about causing similar symptoms. This is a common additive that is found in many things that you would find around the kitchen such as ketchup, mustard, baked beans, carbonated beverages, and mayonnaise just to name a few items. If you ever experience symptoms shortly after consuming your dinner, my advice to you is to check the ingredient labels on the food you recently consumed to see if there is anything that you might be allergic to. If you start to notice a pattern, stop consuming what it was that made you sick.

The only way that you would know if a product has this is if you were to read the ingredient label. The effect of these synthetic additives on the brain is that they excite the brain cells to the point of killing them. That could explain why the Bucks were victorious over the Pacers that night beating them 101-96 after being behind for twenty points for the better half of the first half – because of my constant cheering and over-abundant energy. After they made a victorious comeback from being twenty points behind, it was because of my sacrifice and excitement – and a seizure that happened on the way home from the game. We were only one block from home when my consciousness lapsed. However, when my mom came to my side of the car and opened the front door on the passenger side, I was awake, with no headache, as if nothing happened. It was unusual for something like that to happen. It is still too early to tell whether it was due to the holistic remedies that I was taking. Research shows that when a seizure occurs, it is the body's way of healing itself. A seizure is really a symptom of something wrong within the nervous system and it is the body's way of correcting itself back to the way it is supposed to be.

That convinced me that it was the synthetic ingredients in the diet soda that caused my seizure. I will tell you for a fact, this was not the first time I had experienced this. Despite this, there are some people that just never learn. If you happen to consume soft drinks, as I often do, make sure that you avoid the diet soft drinks. They are loaded with synthetic ingredients. They don't even taste that good anyways. I have learned how to tell if the soda that my parents have tried to slip me is diet or regular – and made a habit of pointing out to my parents that I know when a soda they tried to slip under my nose is a diet soda. I have made it a habit to go with my father when my soda is refilled so that he doesn't get a diet soda for me instead of what I request.

"Mom, I know that you don't believe that to be true. But, if you don't believe me, why don't you read *Holistic Parenting* and then maybe you will see that I am not lying. I know for a fact that you have never read the damn book, because you have two copies of the book in brand new condition sitting on the bookshelf in the TV room. Perhaps if you got your priorities in line, you would see the logic that Lynn has with regard to the American Moneymaking Administration."

From personal experience, I have found that drinking orange soda, when I was living in Wisconsin, was enough for the most part, to make the aura disappear. My seizures originate in my stomach, where my digestive system operates. So, the preservatives that were in the mustard and mayonnaise were enough to cause a seizure. By acting on my brain, they ended up short-circuiting my brain and causing a seizure to happen. Until I had a reaction to food, I was unable to comprehend how vaccinations could be indirectly responsible for why I have epilepsy. The preservatives in food and in vaccines, sort of act as a common denominator, attacking the central nervous system. That is how I know that vaccinations are directly responsible for the *cause* of neurological problems sans a head injury.

THE HEALTH BENEFIT OF ANGER

Anger is what provokes change. It makes you realize that perhaps things can and should be different. That is what fueled my desire to learn more about what happened to me as a result of being on Tegretol for fifteen years as a child. For some reason, I felt that it was like God was sending me a message in response to the question that I had asked for so many years. When I had lost my insurance, I immediately started to ask questions that

I never asked before in my life. I hypothesized that I had received my epilepsy shortly after my shots were administered around the time I was a year old. Because of the fact that out of nowhere I had a febrile convulsion, I was wondering why my parents were not questioning why I was diagnosed with temporal lobe epilepsy. Perhaps because it was just easier to remain in denial about what the real *cause* of my epilepsy happens to be from.

Denying that vaccinations were even associated with my seizures, I could not believe that two people who were supposed to be there for me were essentially lying to me. I sensed a vibe of a possible hidden motive – like the fact that they don't want to admit to being wrong for twenty years. Routines are boring. If you haven't figured that out, maybe you should go back and re-read Chapter three where I talk about the transformation that I went through when I experienced the energy associated with making a positive difference for millions of people.

On a number of occasions, my father would ask me what my belief was about why I got epilepsy. I replied, "The vaccinations that I received as a child were the reason behind it." He outright denied the connection between vaccination and disease, saying that since he had all his vaccinations and didn't have epilepsy, that it had to do with the fact that my brain must have been overactive since birth. Now, given that he has never had to take synthetic remedies to keep seizures from occurring, he has no way to understand the relationship. He even outright bragged that he was proud that he had his health. Some people would say that "karma is a bitch." What goes around comes around.

The day of reckoning will come for him, and then he is going to wish he had listened to me. I will be sure to remember this and get a good laugh at him. Take me seriously on this. You may be suffering from a disease that you are not even aware that you have. Until you start taking holistic remedies on a regular basis, which is the essence of preventative medicine that works with your body to help your body heal from disease, you may not even be aware that you are at a high risk for developing cancer. I am already screened from getting cancer after starting on holistic remedies.

When, out of nowhere, you find out that your bone density is fifty percent below normal, because of a medicine that you took for fifteen years to control those seizures, and you lose the health insurance that covers your medication, you start to look around for something that caused the problem. As it turned out, had I never been vaccinated, I would not have had a problem with epilepsy – and I personally would not have had a problem with osteoporosis at such a young age. The *cause* of my low bone density was from epilepsy, and the *secondary effect* was a hip fracture.

One would think that it is necessary to be vaccinated against disease. The body has a built in immunity to disease strains. That is why we only get the chicken pox once in a lifetime. Once you get the chicken pox, you are immune to the disease because your body has built up a *natural* immunity to it instead of an *artificial* immunity. A good example of this is seen with women who are pregnant, and are advised to limit their intake of tuna because of the risk of absorption of mercury into the bloodstream. Yet, a lot of the vaccines that were administered in their early inception, had mercury as a preservative. Recently, the Centers for Disease Control took mercury out of the vaccines that are administered after people started to complain about this. However, for a majority of people reading this book, chances are, if you see a holistic doctor, he will tell you that your body has had a good dose of mercury, which has been on record for causing epilepsy and many other neurological problems. However, on a cellular level, vaccines have no place in the medical system.

During the time that West Nile Virus was plaguing our population, there were a handful of people that died from West Nile, but mainly they were older individuals and very young children, who had not built up a tolerance to disease. There were an even higher number of people in the general population that were rushing to get flu shots because of the fear of getting the virus – people that clearly did not have a need to get the vaccination. It was not long, before places were running out of the vaccine because of the panic that was created by the media. This was like winning the lottery in the eyes of the companies that made these vaccinations. Let me state for the record that I was not one of those who was dumb enough to get a flu shot. There was one day when I had to call in sick to my job, but that was partly due to the schedule that I was keeping for myself. All I needed was a little sleep and I was fine.

SOME PEOPLE DON'T NEED A COLLEGE DEGREE

I talk earlier in my book about what I found out about the educational system in America. For the most part, it does not allow people to think for themselves. When I first started out in life, I thought that I was going to be getting a good education by becoming an accountant. I hold an associate degree in accounting, which no doubt, comes in handy when it comes to doing my taxes, which have grown about as complex as the tax code itself. With the amount of money that I am going to be making, I am probably going to have to hire an accountant anyways. Having to make estimated tax payments every three months since I don't have federal, state, and social security withheld was a nighmare – especially since I am responsible for twice the amount of social security that I would normally be held responsible for. That is another reason why people like me are needed. We pay for the social services like unemployment insurance and other social service programs that would not be there if it were not for us. Don't forget that little fact next time you feel like bitching about how much money somebody makes. The amount of money we have to pay in taxes each year pales in comparison to what your obligations are.

But, what education attempts to do is get everybody thinking and not improving. We teach the same bullshit over and over to students in medical school and then wonder why the hell diseases are shooting off the damned chart. In the medical profession, doctors and others are saying that they are looking for the *cause* of all different types of ailments that they come across. My answer is not going to be too popular – but based on my experiences, I can say they are true. Childhood vaccinations, which have as many preservatives if not more, than a lot of the food in grocery stores today, affect the nervous system on a systemic level. If I were not vaccinated, I wouldn't have epilepsy, and I would not have osteoporosis. That is why you hear of a lot of cases involving food poisoning. According to *Holistic Parenting*, 78% of the complaints to the Food and Drug Administration in a given year were in regards to aspartame. You would think that this would prompt them to pull this off the market. Because there is no scientific explanation for it, the FDA does not see a problem associated with the use of aspartame.

This all starts at home in the area of food preparation. When I asked my holistic doctor if there are any foods that I should not consume, he said I should not consume anything that comes out of a microwave. There have been some limited studies on the effects of microwaves and the quality of food. Upon adopting my holistic lifestyle, I stopped using a microwave. I went through an adjustment, such as the fact that it takes longer to heat things in the oven. But from a holistic standpoint, having a microwave serves no purpose. The only time that I am reminded of its absence is when my mother asks me about why I don't use a microwave oven any longer. Other than that, I don't notice its absence.

The ritual of change in our life also means that we are sacrificing relationships that are the life blood of every human being. When you are misunderstood, however, sometimes you would rather just spend your time on your own. Such is the case when I decided to consult with the holistic doctor, who I was told by a number of people to stay away from. The change was almost immediate. My appetite got much better than it was from the time that I was on medicine that had disrupted my normal appetite, my emotions were a lot more stable, as if they weren't better already since I didn't have to deal with my parents, and I was on the remedies long enough to cure me of certain cancers that I may have been predisposed to getting at some point in my life, such as prostate cancer. What holistic remedies do is work with your individual DNA to provide preventative care for life. I am already screened from getting certain cancers.

My pact is to get the truth out about the dangers of vaccination somehow. I don't care if somebody threatens my life. You might want to take a look behind your back if you think about something like that, however.

Chapter 8

My Pact With God

Over the course of an entire year, I had gone through change that was so immense that I knew that life would never, ever be the same for me again. On some level, I knew that this translated into some of the messages that I was getting from the sermons that were given at church. Being told that there was joy through suffering, that meant that even though one was suffering, there was a reason that they suffered. I figured that it was my responsibility to everyone involved to make sure that my story was told. I would not only be found guilty for letting millions of people down, but I would also feel that I had violated my moral obligation to let them know that there was a serious problem that needed to be brought to their attention.

Even for those people who did not have to take any of these medications, my story would be enough to hopefully get the attention of those who could get my story to the silver screen. We are all similar in the respect that we have different experiences in life. But then, some people have experiences so profound and original that they end up changing the world for the better. That is what my attempt is with this chapter. In life, we try to pigeon-hole people so that everybody is the same. Unfortunately, there are some people that this method does not work with.

People go to church for a reason. Usually it is because they need to be in touch with God in some way. I have found through my journey that there are people who are afraid of the message that I have just like there are people who are fearful of The Lord Himself. When it became clear that I had a battle waged, I had to stick to my guns hard regardless of what those in my immediate family thought. It is my pact with God to make sure that my message gets out and that the right thing is done for tens of millions of people who are currently in need of assistance.

For the first time in my life, I was spending all of my time alone. It was through this experience that attracted me to a local church in my area, which thereby guided me in the right direction. My euphoria had grown stronger than it ever had since I had been in Arizona. It was this fire growing within me that was bound to change something, somewhere along the line. I didn't know what to do with this immense power that I held. It was clear that for the first time in my life, I was experiencing something in my life that nobody in my immediate family was able to experience first hand on a regular basis.

My Pact With God

I was alarmed. I had moved myself outside of my familiar network where I had grown up for over twenty years. Used to going out to movies every weekend with my father and my brother, I came to realize that something was going to change – in a very major way. For right now, things seemed to stay the same. I had jobs with temporary agencies to do data entry and accounting clerk jobs, and then something changed. After getting introduced to a church, I had some answers to questions that I did not have before, such as why I was so happy, despite the fact that I had been through a life-altering trauma.

As I sat in the trauma center of the hospital awaiting my surgery, I was unsure of myself. Why am I here? What purpose do I serve? This was the first time that I had ever had to have any kind of a major operation. It was during the time that I was in rehabilitation that I figured out what it was that I was thinking. By trying to do the right thing, and take any job that I could find, which at the time to deliver telephone directories, I would be guided by the proper powers. After doing a lot of thinking, I realized it was not routine for a twenty-six year old to have osteoporosis.

Analyzing why this happened, I realized that things were going to change for the rest of my life as I knew it. Eventually, I would not be working as an accountant, but I would take the role of being a teacher. God is really a teacher in many different ways. He teaches us things that we otherwise would not know about ourselves. Using the knowledge that I had that nobody else had access to, I realized what I needed to do. As I was terminated for the millionth time, because I had too many doctor appointments and they were getting in the way of my work, I threw caution to the winds and started putting my experience down on paper. Having the personal foundation nearby in case I had a question, I realized that despite the fact that my family had a problem with me being elevated above them, I had no problem with putting my foot down.

Having epilepsy was easier. It was easier for me to just have a job, and go to it, staying in a routine. Having to deal with osteoporosis was a challenge. How do you go about telling people – "I fractured my hip in three places a year ago" – and get a blank stare? At the same time that I was different, it was like there was a down side to this difference that was creating a rift between me and everybody else out there.

There were times that I was desperate and looking for a way to get some cash flow. Every time that I went to a job interview, I prayed that by some chance, the person who I was interviewing me had not read about my story. Despite the setbacks that I was going through, I had to remain true to my word and get my story out there. The more that my father tried to make my life difficult by not showing any support, the more that I had to make it clear that things were going forward as planned whether he liked it or not.

Every time that I left my apartment, for the first time I had to deal with rejection, because some people would have a problem with me. Never having the experience of being in the spotlight, I had to have a coach somewhere who would be behind me and my projects one hundred ten percent. It takes a very strong person who ventures forward undeterred by other people's negativity to succeed in this business.

I finally realized something through the euphoria that I was experiencing. I had an asset that somebody wanted – knowledge about why things happen the way that they do. I just knew that somebody out there somewhere would hear the story that I was trying to get across to the world and see the valid purpose in it. The only problem, was that I could not get a voice in this world without the help of someone else who believed in my story. My pact with God is that I am telling the honest truth about what happened to me on that fateful but joyful afternoon on April 2, 2002. It was the day that everything in my life would eventually change the entire world forever as myself and everybody would know it. It would also give me a completely new responsibility to be a teacher. I never thought that teaching would be such a respected profession.

It finally occurs to me that I have a monopoly. Just like Microsoft, I am probably one of the only person on record who has epilepsy and osteoporosis. Using this as a weapon, I re-establish my priorities in life. Taking time off from school, I pray to God that I finally get somebody to understand what I had been going through. Unfortunately, just as there are people that don't believe in God being a teacher, there are also those who don't believe that having osteoporosis at a young age can make one a teacher to those who have to take anticonvulsant medication. As I was sending out screenplays related to my story and querying producers to request my work, sometimes I would get asked if this was really a true story. It was so unbelievable, and perhaps that is why a lot of people passed on it.

Before I moved to Phoenix, where I had no family, I had to learn a completely different lifestyle. On the weekend, I used to go to movies with my father and brother. I was in a fix. On a routine trip through the mall, stopping in one of my favorite stores, of all places, the software store, I notice somebody who has the same name that I do. Before I knew it, I was introduced to the church that would give a lot of meaning to my life. It would also tell me why I was drawn to Arizona in the first place. My father was the one who told me that I would make a good teacher. I think that he should be careful what he wishes for. What is it about the desert region of Arizona that is putting my mood at ease? Before I moved to this area of the country, I never thought that euphoria would be something that would have a direct connection to my situation. As some research shows, the emotional health is directly linked to your personal health. There would be many times that before I would have a seizure, I would get real emotional. I didn't understand the mechanics behind this. When my father said that I would make a good teacher one day when I was in high school, I rejected the idea on principle. What did I have to teach to the world. He said that I could teach about what it is like to be a person with epilepsy. Ten years later, that becomes a true life reality as I embrace it and finally have the authoritative voice to talk about how low bone density is connected with anticonvulsant medication.

Before this happened to me, I had no way of knowing what a seizure even looked like, because during my seizures, I was unconscious. I had no idea that I would be touched by something many, many years later that would finally give me the voice of a teacher.

On one Sunday morning, the pastor was talking about something along the lines of being able to derive joy through suffering. Gee, how true this is, even with myself. I figure that I must be suffering for the same reason. I am trying to get a message out that nobody seems to want to listen to. With my immediate family treating me like the black sheep of the family, because I was distinguishing myself, this had some relativity to it. That is when I came to a conclusion. There are some people that actually run from the truth. They don't want to hear how they may have been wrong for the longest time and have to live with that for the rest of their lives. They don't want to have to take advice from somebody who shouldn't have this knowledge in the first place. This is the same problem that God has in getting people to listen to what is considered the truth.

My opinion, is that people who do research on what is really truthful should be respected instead of told that they don't know what they are talking about. My pact with God is to show that there is a better way to do things in terms of medical care. Who ever said that one way of doing things was right? Some people say that holistic medicine is quackery because of the fact that there is no scientific basis behind why holistic medicine works for one person. Hooked into a person's DNA, the holistic remedies help the body with whatever ailments are present, whether it is an appetite that is needing to be fine-tuned, or whether the remedies work to help clear disease out of a person's body forever.

AN UNFULFILLED PROMISE

The pastor was talking about how The Lord was put through suffering because he had a prophecy that he had yet to fulfill. Relating that to my own experience, I thought about the blank stares that I would get as I would tell people about my malady – and that is when I got buried in a ton of bricks. I realized that the same system that was trying to help my situation, was at the same time, screwing it up. I could only imagine what would be coming next. As painful as it was, I had no choice but to find a way to overturn a system that had been in place for well over a century. It was definitely not going to be easy.

When I talk with my friend, Lynn, we periodically get to talking about how the American system of medicine is screwed up in this country. It is the power of my euphoria that keeps the passion of fire alive within me – while I look for a solution to the madness that my father denies. If a lot of doctors don't know how anticonvulsant medication is supposed to interact with the body, the only person who can guide me through that is God Himself.

Being alone has also afforded me time to really decide how much I want to listen to those around me, who are unable to understand me on any level whatsoever. While my family is doing the same things they always were doing, I intend on becoming a teacher to those who need to hear my message. It seemed like my parents were being unsupportive at all costs whatsoever, by saying "You always used to get jobs, Eliot." Yes, but that was until my hip got fractured. Pretty soon they are going to wake up – when I engulf their life in a fire that they are going to have no choice except to let me be who I am and follow my heart through telling this story to the public.

I enjoy fires – especially when I am the one setting them. It gives me a sense of power. Unfortunately, there are a lot of people in my immediate family that keep confiscating the lighter.

Chapter 9

The Vitamin D Epidemic

According to statistics from The National Osteoporosis Foundation of America, there are 34 million people in the United States who are at risk for developing osteoporosis. In addition to that, there are 10 million Americans that have osteoporosis. Out of these 10 million, 8 million are women. Due to the fact that men produce testosterone their entire life, they are at a lower risk for developing osteoporosis. A lot of the women who get osteoporosis are past the age of menopause and, due to bone softening, are at an increased risk for bone fracture.

When I was growing up, bone density tests were not a standard procedure. Doctors had no idea that a lot of the earlier medications prevented the body from being able to absorb vitamin D properly. Vitamin D, a critical component for building strong bones, is made when the body is exposed to the ultraviolet rays of the sun for a short period of time. Known as the sunshine vitamin, this is metabolized in the liver – the same part of the body where the anticonvulsant medication is processed.

However, the lack of vitamin D does not just affect a person's bone health. Research has found that it also has an effect on the emotional balance of a person, especially if they have seasonal affective disorder. Especially for those who have epilepsy, people who have frontal lobe epilepsy have a change in their emotions shortly before going into a seizure. It is said that being happy is the best anticonvulsant medication for those with seizures.

The research showed that a lot of the earlier anticonvulsants such as Tegretol, Dilantin, Depakote, and a handful of other medications played a role in preventing the body from building strong bones. Not getting enough vitamin D can exacerbate osteoporosis.

The National Osteoporosis Foundation reports that in 2001, the number of fractures associated with osteoporosis was reported to be 17 billion dollars, which comes out to 47 million dollars every day. Like everything associated with medical care, don't count on that number getting smaller anytime soon. It will only get bigger, until the right people can do something to fix the problem.

Another way of describing osteoporosis is as porous bone. It means that holes are forming in the bones, similar to the holes that are in a sponge. It is known as the silent killer because a person can have osteoporosis for many years and not notice any major symptoms. It is a major health concern for all involved. Women who have epilepsy have to especially take note, because they are the most likely to get osteoporosis from medication.

According to a study that was done in *The Reader's Digest,* they made the claim that a lot of people in this country are not getting enough vitamin D, also known as the sunshine vitamin. This vitamin is necessary to build strong bones so that then fractures are prevented. The vitamin D, which attaches itself to calcium, is what produces strong bones. According to this study, there are only a few regions of the country where there is direct sunlight year round. In most regions, there are only four months out of the year when there is no direct sunlight. The body needs the ultraviolet light to produce vitamin D in the same way that plants need the sunlight to get the process of photosynthesis kicked into full gear. Just a few minutes out in the sun each week will give the body enough vitamin D, which is generated in the liver, to help the body. Research has also shown that vitamin D has other health benefits besides stronger bones.

The diagram on the left shows the regions that get direct sunlight all year round, and which regions get direct sunlight during which times of the year. Each region is different in terms of which months they get direct sunlight. Generally, most of Texas, the part of California below Los Angeles, and the panhandle of Florida get direct sunlight during all twelve months of the year. If you don't live in these areas, you have to either take vitamin supplements, or you have to get your vitamin D from other food sources such as seafood or tuna. The most effective, however, is to get your vitamin D through direct sunlight. I don't personally recommend you use tanning beds. That is artificial light and is not good for the body.

All those kids have a point when they decide to take off for Miami or some other hot spot in southern Florida during spring break.

ULTRAVIOLET LIGHT

The amount of Vitamin D that your body can make depends entirely on the region of the country where you live. As you look at the diagram, you will notice that in a lot of cities in the United States, your body cannot produce vitamin D for four months out of the year. It is more complicated than that, however. In a lot of cases, just having direct sunlight is not enough to produce vitamin D, especially if you are taking medication that prevents your body from being able to properly metabolize the vitamin D that it receives from ultraviolet light. The way in which ultraviolet light makes vitamin D is similar to the way that the sun makes a plant produce sugar through photosynthesis. The plant has already made the sugar, but without the fuel from the sun, it is unable to produce the sugar, which enables it to grow to a larger size.

The areas that are above the 35 degrees of latitude, which include the areas north of Los Angeles in the state of California don't get direct sunlight from November through February. This is the time when some people get affected by seasonal affective disorder, which is a change in your emotional state based on the amount of sunlight your body receives. I read some research that showed that vitamin D is not only good for bone health, but also for our emotional health. When I was growing up as a child, my body reacted in various ways depending on the weather. Anytime I was in a bad mood, that was how my parents knew that I was going to have a seizure. This also seemed to happen anytime it was cloudy outside, which was pretty much all the time based on the fact that I used to live in Wisconsin, where the cloudy weather seemed to roll through on a regular basis.

When I first visited Arizona, I immediately noticed a difference in my mood because of the weather, and thought it had to do with my vacation. Now, if you look at the diagram on the previous page, you will notice that the entire southern part of the country gets direct sunlight for twelve months out of the entire year. The body has vitamin D receptors in different places and not only helps bone health, but it also affects emotional health. Then, I drive back home to Wisconsin, and I remember it was raining the night that I got back in. It didn't take too long for my body to react to such a change, resulting in depression. This is similar to what people experience who are affected by SAD (Seasonal Affective Disorder), which is why I moved down there two months after getting back and hitting mass depression again. The areas that get direct sunlight year round are central and southern Texas, the southern states along the Gulf of Mexico and the southern section of Arizona. Arizona gets more than 300 days of sunshine during the year. As my body would grow to depend on that, sometimes when that was interrupted, I would have a seizure simply because of the barometric pressure change which precedes a thunderstorm. The change in the barometric pressure affected the fluid inside of my brain, which was in the general area where my seizures occur. Since my seizure activity is located in the area of the brain that involves my emotions, that is why my parents would watch my emotions and use that to gauge when I should be taking a break.

WHAT THEY DON'T TELL YOU ABOUT SUNSCREEN

Even if you live in an area that gets a lot of sunlight, the whole purpose gets defeated when you apply sunscreen as a way of preventing skin cancer. When sunscreen is applied to the skin, that deflects any ultraviolet light from hitting the skin. Generally, all you need is about ten to fifteen minutes per week of exposure to the sun and your body should be able to produce the vitamin D that it needs for strong bones. We all grew up listening to our parents too much instead of what we should have been doing. Anytime we would be outside for a long period of time, our parents always made us put sunscreen on to prevent skin cancer. I stopped using sunscreen.

There are unhealthy elements in sunscreen that are simply not good for the skin. The research in *The Reader's Digest* has shown that even using a sunscreen with an SPF factor of 8 was powerful enough to block the ultraviolet rays from hitting the skin, and our body does not get the vitamin D supply that it very much needs. We may be preventing skin cancer from using sunscreen, but at the same time, we are putting our body at risk for other types of cancers. There are vitamin D receptors all over the body and vitamin D has been shown to benefit the kidneys, intestinal track, and our emotions. Every age has a certain number of IU or international units of vitamin D that they are supposed to consume every day either through food, or through a vitamin supplement.

OSTEOPOROSIS AS IT RELATES TO EPILEPSY

When I was on anticonvulsant medication in my teen years, I was never once given a bone density test. The doctors didn't even know that this was an issue until around the time that I was having hip surgery. That explains why doctors were unable to appropriately diagnose my condition. According to a research article in *The New England Journal of Medicine,* a sample study was done on the effects of anticonvulsant medication on a sample group of 81 children and adults. The results showed that 59 percent of the participants had osteopenia, which is enough loss of bone which would show up on a bone density test. Out of the adults that had a bone density reading, they were 32, 47, and 52 years of age and had epilepsy since childhood. The children who participated in the survey showed no change in the bone density. At this point, it is unknown why there was no noticeable change in children who take anticonvulsant medication.

With regard to teenagers, there was no evidence of bone softening. This happened a lot more with the earlier enzyme-releasing anticonvulsant medications such as Dilantin, Tegretol, Phenobarbitol, and Depakote. The enzyme releasing component of these medications has to do with the fact that these medications are processed in the liver, the same part of the body that manufactures vitamin D. This kept my bones from fully developing. Every once in a while, I would have to wear a heal cup because of pain that I would have periodically as I was younger. I would also notice this when I was rollerblading on a regular basis. The results of this test show that just being on medication for six months produces enough of a change in the bone formation. People who are most vulnerable are individuals who were on numerous anticonvulsants to control seizures, and also those who did not get adequate exercise. For those who have had epilepsy for a long time and been on these medications for a decade or longer, there has been sufficient bone loss that is going to come back to haunt them when they least expect it.

When I had my initial bone density reading, it was −4.6, which was outside of the normal range of acceptable readings. The way bone density is read is by looking at your t-score, or the t-statistic. For every percent below 0, that represents ten percent below normal. In my sample, my t-score was −4.6, which meant that my bones were close to fifty percent below normal. I have been taking 1500 milligrams of vitamin D with Actonel for over two years and had a follow up bone density reading. Curious to see how my diagnosis had changed, over the two year period, I went from −4.6 to −1.3, which was a diagnosis of osteopenia, which is still osteoporosis, but a diagnosis that is not as severe as the first bone density reading. Due to the fact that I am constantly producing testosterone, that helped in my recovery. According to my internist, Dr. Frank Metzger, he said that if I were female, that I would probably only see a reading of something in the range of −3. This is due to the fact that females stop producing estrogen shortly after their childbearing years after they hit menopause. So to an extent, they have two problems to work with. The treatment for epilepsy, is what exacerbates osteoporosis. Osteoporosis was long thought to be a woman's problem and something that men did not have to worry about.Estrogen is what keeps bones strong. When that is no longer present in the body, supplements have to replace what the body used to be able to produce. The diagram above shows the age groups and the relative amount of vitamin D.

With men producing testosterone throughout their entire life, osteoporosis is more manageable with men, unless they happen to have low testosterone levels. When I was referred to an endocrinologist to check my hormone levels, the report came back as being inconclusive. What caused my osteoporosis was not any longer in my system. A lot of the drug companies were praying that with the newer medications, this problem with bone marrow depression would be history. What was never disclosed until recently, was the risk for low vitamin D levels in patients who have epilepsy.

If you have not taken a statistics course, you probably would not understand how the diagram below is supposed to read. If you look at the equation, it is the equivalent of a two-sided test. Either a person has bone density that is in the normal range, or they have a bone density that does not fall within the prescribed range.

From a statistical standpoint, in a normal t-distribution, 95 percent of the samples will be clustered around zero for bone density. 2 ½ percent of the sample population will be in the range below 0 in the left side and have osteoporosis and 2 ½ percent of the sample population will be above 0 on the right side. The sample population that has above normal bone density are the ones that have huge muscles and are not as likely to incur a fracture. When I had my bone density retaken in July of 2004, my sample statistic fell in the area that was in the 2 ½ percent below what the normal amount of bone density happens to be. A bone density test is done by comparing how old you are and what your bone density should be in comparison to other people in the population who fit the same criterion that you have. It wouldn't do any good from a statistical standpoint to compare what my bone density is to that of somebody over the age of fifty years of age because of the huge age difference. That would produce inaccurate data because the way the data is compared is to look at the data for where I am relative to where I should be for somebody who is my age. The area where the red star is located is approximately at −1.3, is what my test results turned up. In the statistical equation, it would read something like this:

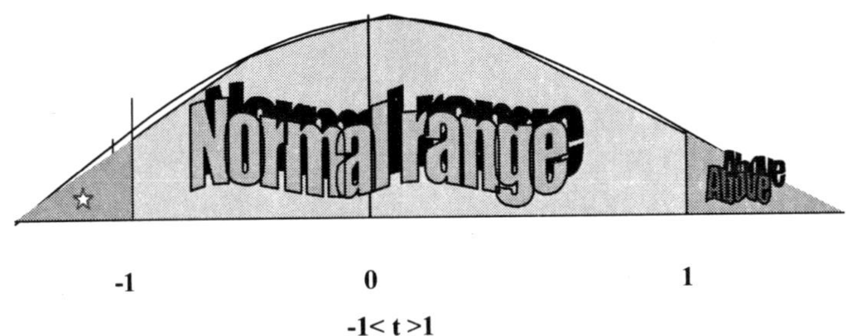

In an algebraic manner, any number that is between these two numbers falls in the normal range. A sample is either greater than –1 or the sample is greater than 1. That is the two-sided test. The number 0 is what falls in between these two tests which indicate a normal bone density reading.

This means that the sample will generally fall in the area of 0 under the bell shaped curve. If you substitute a number into this equation and it fits, then that is a statistic that would fall under the normal bell shaped curve, which represents 95 percent of the sample population. The idea with making the range so small is to be able to effectively treat the people who are at risk for a fracture. If the range were large, there would be no way to detect osteoporosis and get people treated. The people who have normal bone density will have a reading very close to 0. In my case, my reading would not even show up on the chart. From a statistical sense, I would be considered to be an outlier, or a deviation from the normal statistical mean. Bone density in a statistical sense is measured using this t-score. For each point below 0 is a representation of ten percent below normal, assuming that the normal range for 95 percent of the population is at 0. When I had my bone density reading after my surgery, my bone density reading, or t-score, was –4.6, or more than forty percent below normal.

The only way, in a lot of cases, that someone finds out that they have osteoporosis is when they fracture something. According to The National Osteoporosis Foundation of America[1], osteoporosis is responsible for over 1 ½ million fractures every year in the United States.

---

[1] http://www.nof.org/osteoporosis/diseasefacts.htm

Location of Fracture	Number
Spinal Column	700,000
Hip	300,000
Wrist	250,000
Other	300,000
Total	1,550,000 fractures

There is a reason why people get shorter as they get older. It has to do with the bone loss in the spinal column, which is what attributes to roughly half the number of fractures that go undetected every year in the United States. This can be prevented by taking calcium fortified with vitamin D every day and taking Actonel. It won't replace the bone that was lost, but what it does is make your bones stronger to prevent another fracture from occurring.

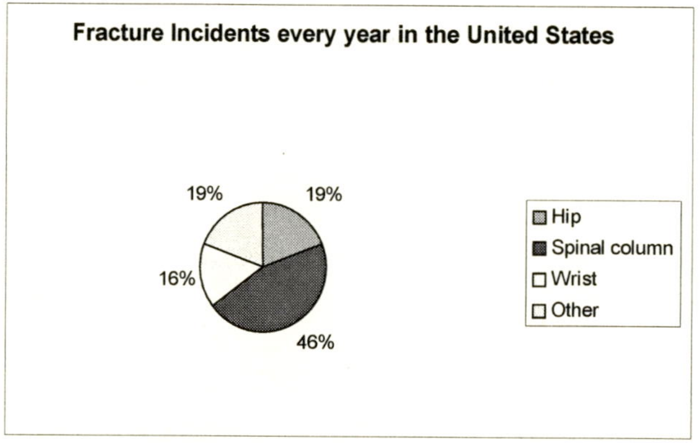

According to an article in the February 1, 2003 issue of *Family Practice News,* there are not many neurologists that regularly screen patients with epilepsy for bone loss. Before medication allowed people with epilepsy to resume a normal lifestyle, a lot of the studies that were done with regard to bone density loss were done on institutionalized patients and people who did not get enough exercise and who were more likely by nature to have a low bone density. But, as epilepsy medications improved and allowed those with epilepsy to have a more normal quality of life, it is slowly becoming an issue that has to be addressed. These same tests were not done on those patients who were ambulatory and got regular exercise. With myself, I got exercise in the form of rollerblading everyday, and was out in the sun a lot. Despite this, my body was blocked from being able to make vitamin D like it should have been able to. But, even today, a lot of neurologists don't do regular screenings for osteoporosis because they don't see the seriousness of the problem.

According to a survey published in 2001, only 27 percent of neurologists did bone density screenings on patients with epilepsy and only 7 percent advised patients to take calcium and vitamin D supplements.[1] When the newer epilepsy medications were made available, specifically Neurontin and a bunch of other anticonvulsant medications, the quality of life for those with epilepsy improved greatly. It is still unknown whether or not the newer medications also cause osteoporosis. Because of this, it is advisable for all patients who are on anticonvulsant medication to get regular bone density screenings for osteoporosis and also to take calcium fortified with vitamin D on a daily basis.

With the majority of people getting epilepsy in their early childhood, there was a good chance that they were on some of these medications. With the newer medications, it was thought that the issue of bone density would not come up as it relates to epilepsy. According to Professor Mattson, "more evidence needs to be looked at before we come to a conclusion." It is unclear whether or not the more modern medications affect the body's ability to produce vitamin D that the earlier medications had. Since a majority of the modern medications are processed in the kidneys instead of the liver, it probably is unlikely for vitamin D loss to be a problem. But if you were diagnosed with epilepsy around the time that the first medications came out to treat epilepsy, chances are you are affected and just don't know about it yet.

Dr. Richard Mattson, professor of neurology at Yale University, says that "reports of fractures and osteoporosis associated with drug treatment of epilepsy date back to the late 1960s. The significance of those reports has not been fully appreciated." According to this same article, it reports that "one of the earliest attempts to explain why epilepsy

---

[1] *Family Practice News.* February 1, 2003

drugs might contribute to bone loss postulated that drugs such as phenytoin (dilantin), phenobarbitol, and carbamazepine (Tegretol), interfere with vitamin D metabolism." There have also been reports that Depakote, which is another earlier medication, played this same role. People who were on many anticonvulsants had a higher chance of getting secondary osteoporosis.

With my age and gender, it was fairly easy for me to hypothesize that medication had something to do with why I ended up in the hospital. But for people who are of the age where menopause takes place, or for those over fifty years of age, the issue of how a fracture occurs can get cloudy, especially if they are taking anticonvulsant medications. Osteoporosis is one of those diseases where you can be living with it for decades and show no symptoms for it until a fracture occurs. Had I not taken that job to deliver telephone directories, I very easily could have gone for a long time without knowing why I was having back pain. It actually all started back in childhood when I had an annoying pain in my heel after rollerblading . I had no idea that I was slowly losing bone mass. As a general rule, children do not show any noticeable signs of bone loss. After they are in their mid twenties, they are more vulnerable to fracture, since the peak bone mass for males is around twenty-five years of age.

With this new information, anybody who is taking anticonvulsant medication needs to be aware of the risks associated with fracture, particularly later in life when the bone density has reached its peak. It is advisable for those who have taken anticonvulsant medication, particularly any of the earlier medications that were enzyme inducing, to be screened for osteoporosis.

Chapter 10

Demonstrating Market Initiative

Despite my injury, it became clear to me that even when I had a website up that talked about my injury in detail, it would be difficult to convince a movie studio of the marketability of this idea. This was something that was never heard of before. If doctors were hearing this for the first time, think about what I was listening to, as I am told by numerous sources that there is no market that exists for my movie.

What is really being told to me, is that they don't want to take a risk on something that has never been done before. I knew that the numbers were there. People all over America were interested in my story and why I had a fracture when I tried to lift some heavy telephone directories. Putting them in the trunk of my car, I realized for the first time that I had a problem that I was going to be unable to describe to people. It would get even more complicated when I tried to sell this idea to Hollywood and be told that the idea was not commercial.

Using the forces within me, I tried to figure out a way to get some people in the movie industry to take my story seriously. It proved to be a very difficult battle. After all, osteoporosis is something that is thought to only affect women who are past the age of menopause. Nobody had ever heard of young people getting osteoporosis, or for that matter getting osteoporosis from long term use of anticonvulsant medications. That is when I took a very bold step and decided that the only way to get people to listen was to get this book published, even if that meant that I was initially going to have to self-publish the book until I could get an agent or publisher to listen to my story.

Sitting in front of Lake Michigan with my mother, she would always point out the faint horizon, equating that to hope. Every time we went for a walk, we would go over there as she would point out that things do change. I knew that things were going to change. I was going to make them change. The power and charisma behind my story would be enough to eventually land me a high six figure contract with a production company, freeing me from the four walls that were the equivalent of a prison. I needed out, and I needed out very soon. I was starting to go stir crazy, counting down the days before I would invariable act on my word and just disappear from existence. My mother was going crazy, as my unemployment benefits had expired, and every bill that came through my mailbox, I had to have her pay – out of her paycheck. What she never told me was that she was about to lose her job.

My dad took the liberty of telling me that I needed to get a job. "Thanks, dad. Like I haven't heard that about one hundred billion times from you already. I am doing everything that I can do. It doesn't help that we are at war right now with Iraq.

Having epilepsy and osteoporosis has taught me a lot about life. It has taught me a lot about relationships. Sometimes there was so much conflict that I thought that I would explode in anger – especially with my dad, who was getting annoyed at the fact that this had disrupted my income. At the same time, he did not want to help with getting things off the ground. I had the data to show that things would change for the better. He wanted me to get a job, slaving away at something that I didn't care about. Why? Because he was in complete denial about the truth with regard to vaccinations being related to disease. Both my mother and father, although supportive, had been making complete pests of themselves as they told me to get a job.

I could look for jobs that didn't even exist until I was blue in the face. Or, I could do the work that I tend to enjoy – the work that I was intended to be doing. Writing a book about the link between epilepsy and osteoporosis and the experiences that occurred as a result. I was going to end up on the street pretty soon. With push coming to shove, no money coming in, and nobody believing my story – despite the fact that people in droves wanted to hear about what I was experiencing as I had titanium pins put into my hip so that I could walk, I put my ego aside and self-published this book.

The trip that my mom was planning to take to Europe two years ago, was delayed not only because she had to help me financially, but because of the trauma that I went through, royally screwing everything up. That is another thing that I planned on doing someday – booking a trip to Europe. My mom had been taking Norwegian classes, she was literally trying to speak to me in the Nordic tongue – and I stood there like an idiot, not even understanding a word she was saying. I guess that sort of parallels the way I was educating myself about Hollywood, a world that is completely unfamiliar to her. Regularly, she would check out books which showed pictures of Norway – the only country in the world to have great fjords, which are land formations which jet out from the coast.

By the time you start reading this chapter, I may already be enjoying myself on the beaches of Norway – a way to make up what was taken from my mom as a result of my relocation and my hip fracture.

Around the time that I was attempting to get my screenplay sold, I was going through an identity crisis. I had been going to Devry University to complete a four year degree in accounting. I already had an associate degree and needed to have a source of income until my book sold. At this point, I was questioning what the hell I was doing. I was trying to resist the process of change. My parents wanted me to find a job, because they thought that what I was telling them for the past two months, were part of dreams that I was having. Grasping at straws, I looked for any indication that something might change.

People knock on the doors of Hollywood every day. They come up with a great idea – and then try to get somebody to buy it. The fact is, after you sell your first screenplay and get "appropriate" representation, it makes it a lot easier to get your next screenplay developed.

Most people who get themselves involved with screenwriting, take up to five years to perfect their craft. We all would like to make a sale, at least in the beginning of 2004, after I had been querying dozens of producers and managers, I finally got somebody to read my script. It was passed on – because of the fact that the script was not properly formatted according to professional writing standards. I am sure that nobody makes a sale the first time through – for any number of reasons. As a result of this experience, I figured out what I was doing wrong and fixed it. On the other hand, I had to be careful of people who were too anxious to screw up good material – only to make it easier to market to a huge number of people.

In a lot of cases, life does not make things happen so quickly. It takes a unique story, personal experience, and knowing that it will be almost impossible to find someone to take your place that gets you working with a manager planning a two or three year contract. What it involves is a story that nobody can turn down. What I never knew, was because I was selling a unique story, I was also put in the thankless position of proving that a market exists for the properties that I am selling. After three months, when the studio wrote back to me through email and said there was no market for my material, I knew the day of reckoning was right on my doorstep.

Difference is what makes this world go round – and it is the same thing that causes turmoil in one's family. At the time, I was trying to pass the excitement that I had for my success onto the rest of my family, with no success. It seemed that my parents disapproved of my choice of a profession.

Hell, my own friends didn't even understand me. Isolated, I had paid a price for fame. By isolating everybody around me based on my differences, I had difficulty getting people to understand me. I told my dad, when I came home for my ten year high school reunion, that there was a good chance that my bank account, which was literally flat broke with $0 in it, would be fattened with an option check from the production company. (By the way, that never happened. The company passed after three months of passing my script around the studio.)

His response was, "The way your bank account is going to get fatter is for you to get a job." I'm going to let you in on a little secret of mine. The ultimate key to success is to break all the rules. Surpass everybody's expectations of who you are going to be as a person.

In the wake of all this happening to me, I found that I was having a lot of trouble getting people to understand my story, much less buy it. I found that I had to come up with a clear marketing objective or I was literally going to be going into Chapter seven bankruptcy. At the time, I had no job, and didn't know how long things were going to take. Just a few months earlier, I had been at my high school reunion, and luckily had found someone who was sympathetic to my story. I was really stuck between a rock and a hard place. My mother had just lost her job. I didn't have a job. I didn't know how long it was going to be until I was going to be out on the street. I was figuring it would be easier if I were dead. I started to slide into a state of depression, which is something that usually does not hit me. Never mind that we have a president in office that doesn't give a shit about the country, which is why I don't have a job, and why I can't find a job.

I guess that is why I have not been calling my parents anymore. Anytime I would call, the subject of money always would invariable come up and I was sick and tired of dealing with it. So, my way of dealing with that problem is to avoid the problem. It's not that I don't want to work. It is simply that it is difficult to find work when you don't have a strong background, not to mention when people simply don't want to understand you because you are different.

And when you just shot your foot off, with everybody hearing about your story, nobody is sympathetic to the fact that I was unemployed for over a year outside of my own control. When everybody expects that you are going to be the same, anybody who is different is looked at as being suspect and a liability. Even if I were to interview for jobs, nobody would want to hire me..

THE MARKETING PLAN

This entire book is the marketing blueprint for the movie script that I had written for Hollywood, the one that they said there was not a viable market for. I would like to give it a year or two and see who is right based on the sales figures for this book. This world is full of people who are different in one way or another.

What I have found, is that difference is what makes this world go round. I have found that going to school and getting a job was an attempt to dumb down my creativity and divert my attention from what I normally would do everyday. Routines are boring. I don't want to stay in a routine. This life experience that I have pointed out in this book should show not only the people who are affected by epilepsy that there is a danger in the long term use of anticonvulsant medication, but that it is futile to change a person once their life becomes threatened by the silent killer.

I don't get inspiration from the lake. I get inspiration from the mountains. At least you can move the mountains to get change started.

Appendix

The following information below is probably most beneficial for those who either have epilepsy or know somebody with epilepsy and are provided for the purpose of putting all the necessary information into one place. I have included some resources as they relate to epilepsy and insurance coverage, particularly about the states that have state-funded high risk pools. Not all states have high risk pools, but the ones that do have risk pools, have to take you even if you have a medical condition. Some of them only cover you for certain conditions. The requirements for getting insurance through one of these pools is that you have to show that you have been denied insurance in the private market due to a medical condition and that you are not eligible for employer-sponsored coverage.

There are only 31 states that have high risk pools with varying waiting periods which mean that you have to be insured for that length of time before you can submit a claim. This information can change at any time, so it is best to call or go on their website to see what the policies for enrollment are for the state where you reside. The information about each insurance carrier can change at any given time, so it is best to check out their website for further information, for those organizations that have websites set up. The following is a list of the information about the states that have high risk pools[1]:

State	High Risk Pool	Phone Number	Waiting Period	Website
Alabama	Alabama Health Insurance Plan	(877) 619-2447 (334) 833-5907	None	www.selb.state.al.
Alaska	Alaska Comprehensive Health Insurance Association	(888) 290-0616 (907) 269-7900	6 months	www.achia.com
Arkansas	Arkansas Comprehensive Health Insurance Plan	(501) 378-2979	6 months	
California	CA Major Risk Medical Insurance Program	(800) 289-6574	3 months	
Colorado	Colorado Uninsurable Health Insurance Plan	(800) 672-8447 (303) 863-1960	6 months	www.covercolora
Connecticut	Connecticut Health Reinsurance Association	(800) 842-0004	12 months	

---

[1] http://www.cobrahealth.com/statehighriskpools.html

Florida	Florida Comprehensive Health Association	(850) 309-1200	Closed to new enrollment effective 6/30/91	
Illinois	Illinois Comprehensive Health Insurance Plan	(217) 782-6333 (217) 782-6410 (TTY)	6 months	
Indiana	Indiana Comprehensive Health Insurance Association	(317) 614-2133 (800) 552-7921	3 months	www.onlinehealthplan.c m/oasys
Iowa	Iowa Comprehensive Health Association	(515) 248-2186	6 months	www.onlinehealthplan.c m/oasys
Kansas	Kansas Uninsurable Health Insurance Plan	(800) 290-1368 x19 (785) 296-3071	3 months	
Kentucky	Kentucky Access	(502) 573-1026	12 months	www.onlinehealthplan.c m
Louisiana	Louisiana Health Plan	(225) 926-6245 (800) 736-0947	6 months Currently there is also a 3 month wait list	www.lahealthplan.org
Maryland	Maryland Health Insurance Plan	(866) 780-7105	1 month	www.marylandhealthins ranceplan.state.md.us
Minnesota	Minnesota Comprehensive Health Association	(952) 593-9609	6 months	www.mchamn.com
Mississippi	Mississippi Comprehensive Health Insurance Risk Pool Association	(888) 820-9400 (601) 362-0799	6 months	www.doi.state.ms.us
Missouri	Missouri Health Insurance Pool	(816) 395-2583 (800) 843-6447	12 months	www.mhip.org
Montana	Montana Comprehensive Health Association	(800) 447-7828	12 months	www.mthealth.org
Nebraska	Nebraska Comprehensive Health Insurance Pool	(402) 343-3337	6 months	www.state.ne.us/home/ DOI/brochure/b_chip.h m
New Hampshire	New Hampshire High Risk Health Insurance Pool	(800) 578-3272 (603) 271-0239	9 months	www.nhhealthplan.org
New Mexico	New Mexico Comprehensive Health Insurance Pool	(505) 816-4248	6 months	
North Dakota	Comprehensive Health Association of North Dakota	(800) 737-0016 (701) 277-2271	6 months	
Oklahoma	Oklahoma Health Insurance High Risk Pool	(913) 362-0040	12 months	
Oregon	Oregon Medical Insurance Pool	(800) 848-7280	6 months	www.omip.state.or.us
South Carolina	South Carolina Health Insurance Pool	(800) 868-2500 x42757	6 months	

State	High Risk Pool	Phone Number	Waiting Period	Website
South Dakota	South Dakota Risk Pool	(605) 773-3148	None	http://www.state p/riskpool.htm
Tennessee	Tennessee Comprehensive Insurance Pool	(800) 669-1851	None	www.state.tn.us/
Texas	Texas Health Insurance Risk Pool	(888) 398-3927	12 months	www.txhealthpo
Utah	Utah Comprehensive Health Insurance Pool	(866) 880-8494	6 months	
Washington	Washington State Insurance Risk Sharing Pool	(800) 877-5187	6 months	http://www.onlir lan.com/oasys
Wisconsin	Wisconsin Health Insurance Risk Sharing Plan	(800) 828-4777 (608) 266-2833	6 months	www.dhfs.state.v sp/index.htm
Wyoming	Wyoming Health Insurance Pool	(800) 442-2376 (307) 634-1393	12 months	

DRIVING LAWS BY STATE

Each state has different driving regulations with regard to the length of the seizure free period that is required before a person can apply for a driver's license. A lot of states have relaxed the waiting period required for a person to be seizure free before applying for a license. A person with epilepsy has to have their physician or neurologist sign something saying that the patient has been seizure free for the required length of time. Depending on the state, there may also be times when updates are required.

Each state has different rules and regulations as they pertain to the seizure free period and whether doctors are required to report epilepsy to the Department of Motor Vehicles.

State	Seizure-Free Period	Periodic Updates	Appeal of Denial	Physician Reporting	DMV Website
Alabama	6 months	Every 6 months until seizure free	Yes	No	http://www.dps.state.a
Alaska	6 months	None unless evidence of an issue	30 days	No	http://www.state.ak.us

izona	3 months	DMV discretion	15 days	No	http://www.dot.state.az.us
kansas	1 year	None	No	DMV discretion	http://www.arkansas.gov
lifornia	1 year	DMV discretion	Yes	Yes	http://www.dmv.ca.gov
lorado	2 years	DMV and doctor discretion	Yes	Yes	http://www.mv.state.co.us
nnecticut	3 months	Based on physician and medical advisory board	Yes	No	http://www.ct.gov
elaware	Physician discretion	Annual	Yes	Yes	http://www.dmv.de.gov
strict of lumbia	1 year	Yes	No	No	http://dmv.washingtondc.gov
orida	6 months	Case-by-case basis until 2 years seizure free	30 days	No	http://www.hsmv.state.fl.us
eorgia	1 year	1 year	15 days	No	http://www.dmvs.ga.gov
awaii	6 months	None	30 days	No	http://www.state.hi.us
aho	6 months	Doctor recommendation	Yes	No	http://www.itd.idaho.gov
linois	None	Medical Advisory Discretion	Yes	No	http://www.sos.state.il.us
diana	None	Doctor's discretion	Yes	No	http://www.in.gov
wa	6 months	Every 6 months until 2 years seizure free	Yes	No	http://www.iamvd.com
ansas	6 months	Annually	Yes	No	http://www.ksrevenue.org
entucky	3 months	Medical Advisory Board Discretion	20 days	No	http://transportation.ky.gov
	6 months	Annually for 2 years	Yes	No	http://omv.dps.state.ia.us
aine	2 years and off medications for 3 months	4 year for minimal/2 years for mild	Yes	No	http://www.state.me.us
aryland	3 months	Medical Advisory Board	15 days	No	http://www.mva.state.md.us
assachusetts	6 months	Medical Advisory Board	Yes	No	http://www.mass.gov

State					
		Discretion			
Michigan	6 months	Medical Advisory Board Discretion	14 days	No	http://www.michigan.g
Minnesota	6 months until 1 year seizure free; Annually until 4 years seizure free	6 months until 1 year seizure free/ Annually until 4 years seizure free	Yes	No	http://www.dps.state.m
Mississippi	1 year	None	10 days	No	http://www.dps.state.m
Missouri	6 months	At License Renewal	30 days	No	http://www.dor.state.m
Montana	Doctor Discretion	Doctor Discretion	30 days	No	http://www.doj.state.mt
Nebraska	3 months	Doctor's Recommendation	Yes	No	http://www.dmv.state.n
Nevada	3 months	Annually	30 days	Yes	http://www.dmnv.com
New Hampshire	1 year	None	30 days	No	http://nh.gov
New Jersey	1 year	Every 6 months	25 days	Yes	http://www.state.nj.us
New Mexico	1 year	Case-by-case Basis	Yes	No	http://www.state.nm.us
New York	1 year	DMV Discretion	30 days	No	http://www.nydmv.state
North Carolina	1 year	Discretion Based on Medical Report	15 days	No	http://www.ncdot.org
North Dakota	6 months/Occupational License at 3 months	Annually for at least 2 years	10 days	No	http://www.state.nd.us
Ohio	None	Every 6 months until seizure free	30 days	No	http://bmv.ohio.gov
Oklahoma	1 year	Medical Board Discretion	Yes	No	http://www.dps.state.ok
Oregon	2 years	Case-by-case Basis	10 days	Yes	http://www.odot.state.o

riodic dates	Appeal of Denial	Physician Reporting	Appeal of Denial	Physician Reporting	DMV Website
nnsylvania	6 months	Medical Board Discretion	30 days	No	http://www.dmv.state.pa.us
ode Island	18 months	Medical Board Discretion	Yes	No	http://www.dmv.state.ri.us
uth rolina	6 months	6 months, annually thereafter	Yes	No	http://www.scdmvonline.com
uth kota	12 months	Every 6 months until 1 year seizure free	No	No	http://www.state.sd.us
nnessee	Doctor Discretion	Medical Board Discretion	Yes	No	http://www.tennessee.gov
xas	6 months	Medical Board Discretion	30 days	No	http://www.txdps.state.tx.us
ah	3 months	Annually until 5 years seizure free	10 days	No	http://driverlicense.utah.gov
rmont	6 months	Doctor Recommendation	15 days	No	http://www.aot.state.vt.us
rginia	6 months	Medical Board Discretion	Yes	No	http://www.dmv.state.va.us
ashington	6 months	Doctor Recommendation	15 days	No	http://www.doi.wa.gov
est rginia	1 year	Medical Board Discretion	10 days	No	http://www.wdot.com
isconsin	3 months	Every 6 months for 2 years	Yes	No	http://www.dot.state.wi.us
yoming	3 months	Case-by-case basis	20 days	No	http://www.dot.state.wy.us